PLANNING, PROFIT AND INCENTIVES IN THE USSR

VOL. I

THE LIBERMAN DISCUSSION
A New Phase in Soviet Economic Thought

Edited by

Myron E. Sharpe

∫

iasp *International Arts & Sciences Press*
White Plains, New York

Acknowledgments

The material in this collection first appeared in *Problems of Economics*, a journal of translations from Soviet sources, edited by Murray Yanowitch, Associate Professor of Economics at Hofstra University, and Fred Ablin, IASP Managing Editor. The translations were made by William Mandel and the APN Agency and were edited by Mr. Ablin.

Library of Congress Card Catalog No. 66-20464

Contents

THE 1964-65 DISCUSSION

CONTENTS

Introduction

The Liberman Discussion: A New Phase in Soviet Economic Thought was originally intended to be one volume of articles. However, events moved so quickly that by the time we were ready to publish this volume, there was already enough material for a second one. For the Liberman discussion is no longer just a discussion. Government decisions have been made on the basis of the proposals of Liberman and his colleagues (in September and October of 1965) and they are being carried out.

The material in the two volumes at hand can be divided conveniently into two major sections and several minor ones. The first volume is comprised of articles which appeared prior to the Government and Party decisions, while the second volume (entitled *Reform of Soviet Economic Management*) is comprised of speeches, decisions, laws and articles dealing with the application of proposals for the reform of the system of economic management in the USSR. Volume One is subdivided into three sections: *Early Presentation*, in which appear articles written by Professor Liberman in 1955, 1956 and 1959; *The Discussion in 1962*, in which a variety of opinions, pro and con, are publicly aired, and *The 1964-65 Discussion*, in which the earlier exchange is resumed, but in which substantial unanimity of opinion is reflected. The consensus, if I may use that overworked term, is neatly summed up in Premier Alexei Kosygin's speech, as well as in the decisions and laws related to it in Volume Two. Subsequent articles explain the significance of the decisions.

The presentation of articles is arranged chronologically. The articles, as the reader will find, vary in quality. Some are theoretically oriented, while others are oriented toward industrial practice. Some deal with a large range of problems, some with no more than one. There is also a repetition of many points, with different degrees of emphasis. However, all the articles taken together should give the reader a rather good knowledge of the issues involved in the discussion, a feel for the kind of problems confronting Soviet economists, and a grasp of the nature of the debate which took place.

It is clear from the articles that the ideas of more than one economist were involved in this discussion. When we refer to the *Liberman* discussion, we do not mean that his ideas were the only ones being discussed. Rather, we are using a convenient label, justified by the fact that Professor Liberman saw very clearly what needed to be done rather early in the game, and stuck quite consistently to his main points until his opinions prevailed. Other economic reformers made no less significant contributions. First the name of the late Professor V. S. Nemchinov, eminent mathematician-economist, comes to mind. Leading roles were also played by Professor V. Trapeznikov, a founder of Soviet automation, and Professor L. Gatovskii, editor of *Voprosy ekonomiki*. Many other economists made important points in the discussion.

For the reader who wants to become acquainted immediately with the essential propositions in the debate, I would recommend the study of the following articles first: *Plan, Profits, Bonuses* by Liberman; *For Flexible Economic Management of Enterprises* by Trapeznikov; *The Plan and Methods of Economic Management* by Leont'ev; *Once Again on the Plan, Profits and Bonuses* by Liberman; and the speech by Premier Alexei Kosygin (in Volume Two). These articles give a rather straightforward exposition of the issues.

What, in fact, are the issues involved in this discussion? First, it is necessary to say something about the historical setting. The Soviet Revolution of 1917 took place in a relatively underdeveloped country which found itself living in a markedly hostile world. These circumstances constitute the original sin which has left an imprint on everything that has happened since. The Soviet Government undertook to build up an industrial economy in the shortest possible time. A high degree of centralization was instituted under ministries (or commissariats until 1946) which had supreme authority in their sectors for the entire country. Capital accumulation for heavy industry was viewed as the key problem, and all else was subordinated to this. In each successive five-year period, beginning in 1929, growth in particular areas was held to be decisive, and the best efforts were thrown into meeting each successive goal. Shortages in the implementation of plans were made up at the expense of areas of lower priority. Through these methods, the Soviet Union did indeed build up its basic industry, and became the second industrial power in the world after the United States.

Following the reconstruction period after World War II, the Soviet Union found itself with a complex economy, with many thousands of industrial enterprises connected by extremely intricate interrelations, with a tremendous assortment of products, with a large cadre of skilled managers, scientists, technicians and workers, with increasing demands for quality consumer goods, and in the main, with a relatively sophisticated level of technology. Under these circumstances, the old methods of highly centralized economic management proved to be quite inadequate. The rate of growth of gross national product continued to be high, but began to decline. The volume of production in relation to fixed assets actually did decline (beginning around 1958-60). Enormous numbers of construction projects were begun and were left unfinished for many years at a time. The application of new technology fell behind international standards in many areas, and unwanted consumer goods began to pile up in warehouses because they failed to meet increasing consumer demand for quality products.

In this situation, Professor Liberman and his colleagues made a number of specific proposals which had the following unifying theme: centralized planning was to be combined with maximum initiative on the part of managers and labor at individual enterprises on the basis of rational cost accounting procedures. There was no point in trying to increase the amount of detail which centralized planning had to supervise, because details were multiplying beyond the point where they could be reasonably handled from any centralized location. The central planners should hand down a limited number of general targets for individual enterprises, and the rest should be left to the managers of those enterprises.

How would the enterprise managers and plant workers be stimulated to increase quantity, quality and efficiency of operation? Economic rather than administrative measures should be used. The old administrative method of planning put a brake on the development of production. If a plant exceeded its plan for a given year, the planners would want to give it a higher quota for the following year, in effect penalizing the enterprise for success. A struggle developed between planners and plant managers over the level of the planned quota. Enterprises were encouraged to conceal assets and unused production potential in order to avoid the imposition of over-strenuous plans from above.

This outdated system had to be replaced by one which used economic levers to stimulate maximum output of high quality goods using up-to-date methods of production. To this end long-term norms of profitability should be established for each type of industry, grouped by plants having the same general conditions, and the managers and workers should be rewarded for attaining their norm, and rewarded additionally for exceeding it. Only those working at the particular plant would best know how to increase efficiency and reduce costs. They would have the incentive because part of the profits would redound to them. At the same time, society would also gain, because the other part of the profits would go into the state budget.

Working with a profit index contrasted with the use of physical indices which promoted irrational patterns of production and plant utilization because of inevitable built-in biases. An index in terms of gross output, for instance, encouraged the use of: 1) more expensive materials; 2) maximum amounts of materials; 3) the production of goods which were easiest to make rather than those which were most needed; 4) the production of goods of low quality, since the production of high quality goods was not always compensated for by a corresponding increase in prices.

This method of planning also discouraged the introduction of new technology for the same reason. A firm would be penalized, because, while introducing new techniques, the volume of production would have to be curtailed until the new techniques were mastered, but no objective price adjustment criteria existed which would require the enterprise to be compensated for its trouble and for the temporary loss in gross output. These difficulties were to be remedied by differential profit norms which encouraged efficiency, economy, quality and technological progress, by the introduction of a system of prices which reflected real costs of production and the degree of demand for the product, and by standards which took into account actual sales, not simply output.

A related difficulty encountered in the old planning system had to do with the indirect relations between suppliers and consumers. Directives were given for the production of a certain amount and type of product by the central planning authority which stood between the producers of such products and their consumers, whether industrial or individual. This greatly complicated the system of supply. Enormous quantities of productive assets lay idle at various plants, in fact, were hoarded by plant managers because of difficulties, delays and aggravations in obtaining supplies of various inputs. The new proposal suggested that direct ties be established between producers and consumers, and that production be based on orders rather than on directives. In this way, purchasers could deal with producers from whom they were reasonably certain of obtaining supplies on time, in the right quality and quantity. On the other hand, a plant manager familiar with the capabilities and resources of his plant could bid more realistically for contracts, with the full knowledge of what could be done and in what period of time.

Direct ties between producers and industrial consumers would not be the only form of obtaining supplies. Some staple items would still be ordered directly by the state and standard items would be stocked by wholesale establishments. But the supply system would be based — not on the cumbersome method of calculating in terms of physical units and the assignment of specific items to plants as heretofore — but on financial trade relations. In short, enterprises would, to an increasing extent, buy their raw materials and other supplies in direct dealings with sellers.

Excessive quantities of supplies were kept on hand and immobilized not only because the system of supply was unreliable, but also because capital assets were free goods — there was no charge for their use. Therefore, the plant manager suffered no ill consequences by having excess inventories lying around. The cost accounting principle [khozraschot] in which the operations of the plant were supposed to be covered by receipts from its output, was violated simply because an entire category involved in production, namely, capital assets, fixed and circulating, had no cost. The reformers proposed that a charge should be made for capital so that the plant administration would be encouraged to minimize the amount of capital used in its operations, including the amount of inventory of various items. Only with realistic, not fictitious, cost accounting would the profitability index have meaning in terms of keeping expenses to a minimum.

All of this was connected with yet another consideration

(mentioned earlier), and that is the price system. Because the theory of value was long thought to be inapplicable or only slightly applicable under socialism, insufficient attention was given to the establishment of a more rational system of prices reflecting actual values; that is — in terms of the labor theory of value — reflecting socially necessary labor. Of course, as prices were irrational, it was impossible for a manager to calculate the best choice among several possible options. Nor, for that matter, was it possible to calculate realistically the level of profitability or the relative efficiency of labor versus mechanization of a particular operation. In order to implement all the foregoing proposals, a rational system of prices had to be established.

This, in brief, was the area covered in the Liberman discussion. How the issues were resolved in theory may be seen from Kosygin's speech. The reformers won the day. But how the issues will be resolved in practice will only be seen over the course of the next several years. The Soviet economy is a very big organism. There are legions of bureaucrats who have vested interests in the old ways of doing things. On the other hand, there are also many businesslike men on the scene who are interested in efficiency and who will fight for it.

Perhaps the most important thing about the Liberman discussion is that there was a Liberman discussion. It could not have taken place a decade earlier for well-known reasons. The least that can be said is that a society which can provide a platform for an objective discussion of some of its problems is making progress.

What is the significance of the economic reforms in the USSR? I have heard many opinions expressed, including the following: they represent an important departure from past practices; they are simply an evolution from past practices and are not particularly new; they are the first step in a long process of evolution to new types of economic planning and management; they represent the adoption of capitalist methods; they amount to a temporary thaw, but soon the old methods will be re-introduced. My own opinion is that — given the fixity of centralized decision-making for aggregate categories — the new proposals, now embodied in decisions, represent a significant change in the method of economic planning and management, a growth in the degree of sophistication necessitated by economic development. An advanced economy requires methods different from those of an undeveloped economy. Since the economy cannot go back to the old stage, neither can the methods of running the economy. That the Soviet economy will also continue to develop in the future is a truism, and that economic methods will continue to develop and meet new needs follows. Today there is considerable discussion about computerizing planning. This, no doubt, points the way to the future. The use of computers in a centralized fashion is absolutely essential to the optimization of economic activity. At the same time, economic incentives are essential to free the initiative and enterprise of individuals in whatever capacity they are working in order to realize an optimum program.

As for the convergence of the Soviet system with capitalism, it should be remembered that the word "profit" does not change a sow's ear into a silk purse or a silk purse into a sow's ear, depending on your point of view. Every economic system produces a surplus, in the sense conveyed by the labor theory of value. If it is utilized rationally in a centrally planned economy, this cannot imply that such an economy is *ipso facto* a capitalist one. What is to be avoided is taking past economic dogmas of the Stalin period, or of any period, for that matter, and setting them up as "Marxism." This is a useless form of pedantry, whether one occupies the defending position or the attacking one.

January 1966   Myron E. Sharpe

Early Presentation

E. G. Liberman

Evsei Grigorievich Liberman holds the Chair
of Economics of Engineering at Kharkov State
University, and is a consultant to the Ukrainian
Institute of Economics and to a number of engineering plants in the Kharkov area. Professor
Liberman began questioning the prevailing conception of centralized planning after World War
II. At a meeting of economists in Moscow in
1948, he proposed the use of profits as a means
of gauging enterprise efficiency. The proposal
went unnoticed, but Professor Liberman per-
sisted in collecting data and in developing his
ideas. He presented his doctoral thesis on the
subject in 1956. His papers appeared in *Voprosy ekonomiki* and *Kommunist* in 1955 and
1956, and he expounded his position at conferences and in survey reports to the Soviet
government. Professor Liberman gained promi-
nence in 1962 when his article "Plan, Profits,
Bonuses" appeared in *Pravda*, stimulating the
public discussion which took place that year.
The discussion resumed in 1964 and led to the
economic reforms adopted the following year.

Voprosy ekonomiki, 1955, No. 6

Cost Accounting and Material Encouragement of Industrial Personnel*

E. G. Liberman

Throughout the entire history of Soviet economic development, cost accounting has played a most important part as a method for the planned management of socialist enterprises. As early as the beginning of the rehabilitation period, Lenin pointed out that solid bridges to socialism must be built not on enthusiasm directly but, with the aid of enthusiasm, on personal incentive and cost accounting. Following these Leninist principles, the Communist Party repeatedly examined and solved at its congresses, conferences and plenary sessions the problems involved in the development of cost accounting. The forms for effectuation of cost accounting have been modified and improved in accordance with changing economic conditions.

The measures adopted for further strengthening cost accounting and improving the methods of managing socialist enterprises are bound to promote the growth of production and labor productivity. These measures are intended to strengthen financial control over the enterprises' operation, to reduce and improve the administrative apparatus, and to simplify the forms of planning and evaluating the work of enterprises.

There are still many shortcomings in cost accounting practice. The experience gained in this area by advanced enterprises has not been adequately studied. A listing of facts, examples, and isolated figures is often substituted for a profound analysis of the enterprises' work. Soviet economists are faced with the task of making a serious effort to improve the system and practice of cost accounting.

---

*The article is presented for purposes of discussion.

3

***

Cost accounting is one of those objectively necessary categories
of the socialist economy which are causatively connected with
the commodity-money relations existing in our country, with
the influence of the law of value on production.

Ia. Kronrod disputes this by claiming that cost accounting
is not based on the use of the law of value and that it is not
causatively connected with the operation of this law. According
to Kronrod, cost accounting is "in no way a consequence of the
operation of the law of value under socialism." (1) Elsewhere
he writes: "Cost accounting is characterized in a number of
works as the result and instrument of the law of value. Actually,
although cost accounting is connected with the system of value
relations, this connection . . . is in no way a connection of cause
(value) and effect (cost accounting) . . . " (p. 30).

It is true, of course, that the cost accounting method of
managing socialist enterprises is connected not only with the
law of value, but with the entire economic system of socialism,
reflecting the demands of the basic economic law of socialism — the law of the planned, proportional development of the
national economy — and other economic laws. However, cost
accounting as a method of managing socialist enterprises has
the specific feature that, due to the operation of the law of value,
promotion of a decline in socially necessary outlays of labor and
control over the planned, proportional development of production are realized with the assistance of the value form. Consequently, cost accounting is connected with the law of value
causatively; it is based on the utilization of this law.

Further, according to Kronrod, the fundamental feature,
the basic distinction of cost accounting, is that "each individual
state enterprise effectuates the process of reproduction through
a planned replacement of its material expenditures, the remuneration of its workers and employees according to their labor,
and the creation of accumulations not just out of centralized
funds and the aggregate national product, but out of the value of
precisely that part of the national product which the given enterprise produces" (p. 11). The first thing that strikes the eye here
is the proposition that the covering of enterprise expenditures
and the formation of accumulations occur at the expense of the
value of the product manufactured at the given enterprise. But
there can be no value without the law of value. Consequently,
had there been no law of value, there would not have been this
criterion of cost accounting. This alone refutes the author's

reasoning to the effect that cost accounting is a special form of production relations which is causatively independent of the law of value.

But there is more to the matter than this. Is the principle of covering expenditures and forming accumulations out of the given enterprise's product sufficient for determining cost accounting? Imagine an enterprise which has replaced all its expenditures out of its own receipts and has even obtained a profit above that planned, but has not fulfilled its plan assignment for production costs. This happens rather often when an enterprise fails to adhere to the plan as regards the assortment of goods it is to produce. (2) According to the criterion fixed by Kronrod, such an enterprise meets the requirements of cost accounting, though actually it has not conformed to the planned norms for outlays of social labor and has violated the proportionality of production. Meanwhile, control over the observance of labor outlay norms and the proportionality of production is actually one of the most important functions of cost accounting.

Let us take another example. According to the criterion fixed by Kronrod, the machine-tractor stations are not cost accounting enterprises because their expenditures are not covered out of their receipts but out of state funds. Nevertheless the machine-tractor stations really operate on the cost accounting principle (though it has a specific organizational form here), because all their expenditures are commensurated not only with norms, but also with production results by comparing the planned and the actual cost of tilling a unit of land; their staffs are rewarded financially for saving funds; they have contract relations with collective farms and suppliers. At present it would be inexpedient to have the machine-tractor stations cover their expenditures out of their receipts, not only because they get part of their income in kind, but also because the substantial investments being made in the mechanization of agriculture are incommensurate with their receipts (on a territorial scale).

Cost accounting is a means of stimulating a steady reduction of the outlays of past and living labor specific for the given enterprise and, thereby, of reducing the socially necessary expenditures. Cost accounting also serves as a means of checking that the output of each enterprise corresponds to the needs of society as regards quantity, quality and assortment, and that the goods are put out on time. The stimulation and control are effectuated with the assistance of the value form.

Enterprises operate under different natural conditions; moreover, they are not equally provided with up-to-date equipment. This necessarily leads to deviations of individual labor expenditures from those that are socially necessary, deviations which do not depend on the enterprises' efficiency. Though the state fixes identical prices on identically named products of labor, it plans different production costs for these goods, depending on the conditions under which the enterprises work. Consequently, the problem is not only one of replacing expenditures and forming accumulations at each enterprise out of the price of the goods it has put out, but also of having the production costs of the given enterprise's product correspond to the outlay norms that were fixed with due account of the given enterprise's individual production conditions. Hence it is clear that the profitability of an individual enterprise cannot conform to the average profitability norms for the entire economy and must be checked for correspondence to the profit plan or the profitability norms for a similar group of enterprises. With uniform prices on identically named products of labor, this cannot be controlled solely through the mechanism of replacement of the outlays for the manufacture of the product out of its price, as the enterprises will inevitably realize this replacement in differing degrees and, consequently, will have different amounts of accumulation. That is why it is not only the very fact of the replacement of production outlays, and the formation of accumulations out of the price of the product, but also the quantitative measure or degree of this replacement and accumulation that gives the possibility of estimating the cost accounting activity of the enterprise. And this can be attained only if the expenditures and production results are commensurated in value form. Evidently, the expenditures (production costs), the receipts, and the profit have to be commensurated with the plan or the norms for the given group of enterprises.

***

Further improvement of the forms of rewarding personnel materially, depending on production results, is a most important factor in strengthening cost accounting.

As regards each individual worker, the problem of commensurating the results of labor and its remuneration is solved through wages, especially through its piece-rate and progressive piece-rate forms. This is possible because the extent and effectiveness of every worker's participation in social labor can be checked relatively simply through the quantity and quality of the goods produced. But it is also important to ensure coordination of the work of the entire body of workers at the enterprise, for the workers' efforts will not produce the desired results without it. Cost accounting is called upon to solve this difficult problem. It is employed to estimate the expenditures and results of the labor of the entire enterprise. The value form is used for this purpose. Without making use of this form, it is impossible in the socialist stage to commensurate the outlays of past and living labor and the results of production. Cost accounting is one of the methods by which personal material interest is put at the service of social production as the sole source for satisfying the material needs of all members of society to the greatest possible extent.

In spite of all the successes scored by our industry, there are serious shortcomings in the quality of management of many enterprises; this applies, in particular, to the organization of planning and cost accounting.

In our view, these shortcomings in economic management should be eliminated not by making planning more complicated, more detailed and more centralized, but by developing the economic initiative and independence of enterprises. The plan is the centralized foundation which ensures the necessary rates and proportions in the development of production. Cost accounting must serve as a means for finding the best ways of fulfilling centralized directives on the basis of the active participation of the enterprise staff in the introduction of new equipment and new production processes. Enterprises must be given broader initiative; they must not be bound by petty tutelage and bureaucratic methods of planning from the center. But, on the other hand, the work of enterprises must be kept under strict control, and tendencies toward narrow departmentalism must be suppressed. This problem can be solved by the proper use of the principle of material incentive; it should really reflect a combination of the personal and public interests, guarantee the observance of state interests and, at the same time, simplify control of the observance of state interests without leading to bureaucratic methods of management.

It is from this point of view that we must examine the problem of improving cost accounting as a method of managing socialist state enterprises.

***

With reference to cost accounting within a plant, it is essential to study the experience of the enterprises which have actually (and not just formally) introduced shop-wide and section-wide cost accounting, as well as the work of those which have not succeeded in doing so. We must bring out the factors that have caused the lag and real difficulties in introducing cost accounting and, on this basis, we must work out sufficiently simple forms and methods of intraplant cost accounting that are applicable to the different types of enterprises and shops, methods that will not lead to the excessive growth of the managerial staff and, at the same time, will guarantee effective control over the production of the goods put out.

Intraplant cost accounting will in most cases be practicable where the administrative and technical personnel in shops are given bonuses not only for carrying out output plans, but also for cutting production costs.

The simplicity and efficiency of intraplant cost accounting must, in our opinion, be based on the following principles:

1. Bonuses for shop personnel must depend on, and be paid out of, savings derived from cost reduction.

2. Progressive outlay norms should be kept stable for the shops for not less than a year. The shop personnel must know firmly that any innovation and any reduction of nonproductive personnel will benefit both the state and the shop. If every improvement introduced by the shop leads to the reduction of norms for the next month or quarter, then material incentive for such improvements is undermined. The workers' pay rates are not changed every time they surpass their output norms. Remuneration rates remain unchanged for at least a year, and this serves as a continuous incentive to exceed the output norms. But if this holds true for workers, it is just as true with reference to the shop's executive staff. The shops can and should be given increasing plan assignments from month to month (including the cost reduction assignment), but this must not be attended by changes in the basic pay rates for at least a year. In this way, the incentive for further improving production methods will be strengthened to a greater extent. Under this system, innovation proposals and organizational-technical measures will be introduced into practice at a faster rate, which is actually the main purpose of intraplant cost accounting.

3. The simplicity and quality of planning are determined by the adequate preparation of norms. With well worked-out norms, there is no need to elaborate production-cost plans or complicated production estimates for the shops: it is sufficient that the shop knows the norms for labor outlays and wages for the goods produced, as well as the norms for material expenditures and indirect expenses. The results of the shop's operation

and its bonus fund are determined by a relatively simple comparison of the actual expenditures of the shop and the outlay
norms for the entire actual volume of shop output.

4. The process of determining the results of the shop's
operation should be, in addition, a process of determining the
production costs for the entire plant. A single bookkeeping system, without taking parallel stock of the plant's production costs
and the shop's so-called cost accounting production cost, must
reflect the results of the cost accounting in shops and the plant's
production costs.

Of course, the detailing of these principles with reference
to different types and sizes of shops and enterprises requires
competent work on the part of economists, bookkeepers, tech-
nologists, and norm-setters. The chief administrations and ministries have an important role to play here; there are extensive
possibilities for drawing research institutes and the publishers
of technical materials into cooperation with industry.

As for the cost accounting of enterprises, it is essential to
examine the question of the material incentive for enterprise
personnel.

The whole enterprise is interested in the growth of profitability, as outlays for enlarging the given enterprise are financed
out of its profits. However, this way of utilizing profits is not
connected directly with the personal material interests of the
given enterprise's personnel. The director's fund has greater
significance in this respect. But the role of this fund as a stimu-
lator seems to be declining at present. The director's fund is
fully formed, in accordance with the results of a year's work,
by approximately the second quarter of the following year. Thus,
the payment of bonuses is quite separated in time from the work
for which they are awarded. Many enterprises are deprived of
this fund if they violate just one of the numerous conditions stipulated (even a single item in the output assortment). In the event
that the enterprise is not responsible for the violation, the fund
may be restored only by the minister, with the approval of the
Ministry of Finance. Under these circumstances, many of the
enterprise's personnel see no direct connection between their
efforts to cut production costs and the bonuses coming from the
director's fund.

The fund for giving bonuses to winners of the all-Union socialist emulation is formed out of above-plan profits. However,
the stimulating role of this fund is greatly limited by the fact
that it is irregular in nature and applies to a relatively small

number of enterprises.

Direct material rewards to enterprise executives for the results of their work are forthcoming in accordance with a so-called progressive-bonus system, which is a serious factor in the effort to fulfill the plan every month. It should be pointed out, however, that this system has the drawback that the size of the premium depends solely on the fulfillment and overfulfillment of output plans; there is no connection between bonuses and above-plan cuts in production costs.

I. V. Stalin emphasized that cost accounting is expected to train personnel to be thrifty. Economical management must provide, first and foremost, that the enterprises themselves find all the possibilities for cutting production costs, being materially interested in doing so. Meanwhile, with the existing bonus systems, this material interest operates only after the plan has been approved; it does not function prior to this approval, and sometimes it even operates in the opposite direction. As we know, plan assignments are distributed among enterprises mainly on the basis of the indices they have obtained. Therefore, inefficient enterprises sometimes get comparatively low plan assignments and the burden of plan fulfillment is imposed on the efficient enterprises. As a result, enterprises often attempt to conceal their possibilities so as not to get too high a plan assignment. It happens that even in the process of the fulfillment of current plans, the system of material incentive has a restraining effect as regards the disclosure of production reserves. In this connection, the chief administrations and ministries are faced with a difficult and sometimes unmanageable task: to determine from the center what the enterprise's actual possibilities are and what kind of a plan it should be given. This produces a constant inclination to exaggerate report data and plan calculations. Enterprise plans are often determined in a kind of "contest" between the chief administrations and the enterprises. We know from experience that such planning procedure does not exclude chance in the distribution of plan assignments among enterprises and frequently fails to ensure the use of available production resources. The enterprise itself knows best what actual resources it has for enlarging and improving production, and it will mobilize these resources more rapidly and completely, the more material incentive it has to do it.

It is particularly regrettable that the present incentive system retards the introduction of new equipment and production

processes. Every important new development involves technical risk and requires time for its assimilation. Enterprises
are willing to tackle these difficulties if their personnel are
rewarded for it. But the effect derived from these measures
is, naturally, included in the plan, i.e., the output program is
raised, the material and fuel outlay norms are cut, and the cost
reduction assignment is increased. And since the entire incentive system is based on the overfulfillment of plans, the introduction of new equipment makes it harder and not easier for
the enterprise to achieve the corresponding level of material
encouragement. This explains the fact that enterprise personnel
sometimes hinder, directly or indirectly, the rapid introduction
of scientific proposals, fail to conduct promising experimental
work under the pretext of "struggling for plan fulfillment," and
are often inclined to continue the production of obsolete types
of goods by outdated processes. Unfortunately, the method of
rewarding for plan fulfillment often turns into a method of encouraging technical stagnation.

It seems to us that it is essential to create a situation in
which the enterprises themselves have an interest in receiving
and carrying out a sufficiently strenuous plan, a situation in
which this does not worsen the conditions of their material encouragement but, on the contrary, improves these conditions.

Let us see, from this point of view, how the party deals
with the issue of material incentive in collective farm production.

What is the hectare principle of estimating state deliveries (and, according to the decisions of the January Plenary
Meeting of the CPSU Central Committee, state purchases)?
It is the principle of a stable norm: the assignments of produce
deliveries to the state are not changed annually and are not
increased if the collective farm draws up higher targets for
crop yields and attains them. Therefore, the collective farm
has an interest in introducing improvements, in drawing up high
harvest plans and topping them, because the better the collective
farm's harvest, the more profit it will get. What would happen
if the size of deliveries and purchases depended on the harvest
plan (i.e., the larger the plan, the greater the deliveries)? It
is clear that the collective farm would strive to draft an understated plan. This would produce the same results as did the fixing of procurements in accordance with actual yields. This
practice, condemned by the party, deprived the collective farms
of all incentive to raise yields and led to the shifting of grain

delivery assignments from inefficient farms to those that worked efficiently.

Collective farms are not state enterprises and, of course, we cannot draw exact analogies here. But the net income of state enterprises, whose product is the undivided property of the state, always serves in the Soviet economy as the source for rewarding the workers who have created this income. That is why, in our opinion, the principle of proper stimulation, based on the fulfillment of stable norms, and not only plans, must produce good results. We consider it quite possible to fix differentiated, stable (i.e., functioning for a long term) norms of profitability for groups of approximately identical state enterprises. The enterprises should be granted bonuses on the condition that they have fulfilled their plan, but the size of the bonus should depend on the degree to which the long-term norms are fulfilled. The size of the bonus in percentage of profit will be larger, the more the profitability of the enterprise surpasses the initial, normative level. In other words, it is necessary to fix for every group of approximately similar enterprises, and for a long term (the same term for which wholesale prices are in effect), a minimum limit or norm of profitability and a corresponding minimum rate of allocations from profits to the bonus fund. Everything attained by the enterprise in excess of this level (regardless of the extent of fulfillment of the annual profit plan) will be divided between the state and the bonus fund of the enterprise. But the higher the enterprise's profitability, the greater will be its share of the profits.

Under these conditions, the director and the entire staff of the enterprise will not object to sufficiently high output assignments, as the enlargement of output is the surest way to cut production costs and raise profitability. As we know, some of the shop's expenses and all the overall plant expenses depend very little on the growth of the output volume. For machine-building plants, these expenditures come to about 30% of production costs. Consequently, if output is increased by only 10% in a year, production costs must come down by about 3% under the effect of this factor alone. With an initial profitability norm of 4% (such profitability is planned, on the average, for machine-building enterprises) and with a production-cost cut of 3%, the profitability level will rise to $\frac{4+3}{100-3}100 \approx 7.2\%$, i.e., it will nearly double, and there will be a corresponding progressive increase of the bonus fund. This shows that if incentive is based on a stable profitability norm, the enterprise will have no reason to strive for unjustifiably low output plans. On the contrary, the director will seek a large work load for the enterprise.

If under these conditions the enterprise itself drafts the plan for production costs (and consequently for material and labor outlays) and profits, then this plan will never be lower than the stable limit; it will take into consideration the disclosure of all reserves, since this will now profit the enterprise directly. No matter how high the plan, the size of bonuses will depend not on the plan, but on the basic norm. And if a sufficiently intensive plan results in a considerable reduction of production costs and a rise in profitability, the enterprise will get large sums for its bonus fund. Of course, these sums will be paid into the bonus fund only if the minimum yearly assignment for output and assortment is carried out. Under these conditions the enterprise will strive to draft a maximum and, at the same time, realistic plan for production costs and profits, one that will obtain the largest possible allocations for its bonus fund. Moreover, the entire staff of the enterprise will strive to carry out and exceed this plan so as to get these high allocations. The incentive to exceed the plans remains, since the larger the profits, the larger the bonus fund. But the "stimulus to get unjustifiably low plans" will disappear because, under the above conditions, the smaller the output plan, the harder it will be for the enterprise to get large allocations to the bonus fund. It must be remembered that output can be increased only if there is a high plan assignment, for that is the only way to get the necessary supplies of metal, fuel, etc. Finally, the staff will know that if they bring out all the reserves available in the given year, this will not raise the initial profitability norm underlying the estimation of the enterprise's success in achieving economies during the next year. It will also be profitable to start putting out new goods and to fill difficult orders, for the price of new goods in the first year (first series) will equal the report calculation checked by the party that has placed the order, plus the profitability rate planned by the enterprise itself. Thus the filling of the order guarantees a planned profitability level and, consequently, a proper bonus fund.

The long-term profitability norm gives greater scope to the introduction of new production equipment and processes. Regardless of the period within which the new production methods start paying and of the extent to which they will raise the plan assignments, the application of these methods will, in the end, produce a level of profitability that is higher than the initial stable level. Consequently, a considerable share of the effect produced by the introduction of new machines and processes will go to remunerate the efforts of the advanced collectives and will encourage them to introduce progressive production methods.

This question naturally arises: how can long-term assignments or norms of profitability be fixed for groups of enterprises if their equipment levels, output assortments, etc., undergo changes? Experience has shown that these changes usually have little effect on the type of an operating enterprise. That is why the diversity of the results of work is not so much due to the difference in enterprise type, as to the difference in the quality of work of the individual enterprises. So, if we give enterprises profitability norms that are on the level of the leading plants in each group, we shall not make a mistake. On the contrary, in our opinion this will help lagging enterprises to catch up with the level of the advanced ones, this being stimulated by the powerful factor of material incentive.

With reference to the most complex production branch — machine building — the following calculations may illustrate in principle how the problem can be solved. When wholesale prices are being fixed, the initial profitability in machine building is fixed within the range of 3 to 4%. If we accept it as the minimum profitability level, then for plants that are better equipped or that have mastered their production more successfully, and therefore have lower production costs, this norm will be higher, for instance, 15% and more. For enterprises operating in average conditions, the profitability norm will be approximately 10%. These norms are fixed for every group of enterprises for the period up to the revision of wholesale prices. Of course, if an enterprise is subjected to fundamental technical reconstruction or if its production specialization changes, it must be transferred to a different group, and this will lead to a corresponding change in the norm. If there are enterprises that manufacture particularly specific goods, the groups can be small (profitability norms may even be fixed for a single enterprise).

Further, it is necessary to fix the share of profits that is to be allocated to the bonus fund. Of course, this fund must be in a certain relationship to the wage fund. It is well known that at a given wholesale price level, wages account for a rather stable share of production costs that is characteristic for each group of enterprises. In most machine-building plants it amounts to about 30%, in machine-tool enterprises — 35 to 40%, and in precision instrument enterprises — some 55%. What relationship should be established between the wage fund and the bonus fund? Let us turn to actual data: according to incomplete data, in 1954 the sum total of the incentive funds for plan fulfillment varied, for individual plants, from 4 to 9% of the wage fund; the bonuses paid out of the director's fund, bonuses awarded in the all-Union socialist emulation, and progressive-bonus pay to engineering-technical and office personnel were included in this estimate. In order to give the enterprise personnel greater material incentive, let us increase this actual range in the size of the bonus fund and assume that the minimum scale of bonuses comes to only 2% of the wage fund, and the maximum — to 15%. Evidently, for groups of enterprises with an initial profitability norm of 10% and a wage fund that accounts for 30% of production costs, the rate of allotments to the bonus fund will be $\frac{2 \times 0.30}{10} \times 100 = 6\%$. That is the minimum rate, as it gives the enterprise a bonus fund of only 2% of the total wage fund (this is equal to about 8% of the fund of salaries to engineering-technical and office personnel) if the minimum profitability of 10% that has been fixed for the given group of enterprises is attained. But allocations to the bonus fund should increase progressively as the enterprise raises the profitability of its operation. So, in order to build a scale we still have to determine the maximum percentage of profits to be allocated to the bonus fund. For this purpose, we have to determine the top limit of profitability: with stable wholesale prices, the profitability of comparable output may rise by the end of the five-year period up to about 35% (if the initial profitability is 10% and enterprises cut production costs by 5% annually). On the other hand, we have assumed that the bonus fund must not exceed 15% of the total wage fund. Thus, the maximum rate for the given group of enterprises is $\frac{0.3 \times 15}{35} \times 100 \approx 12.9\%$ of the total profit. This rate will provide the enterprise which has attained the 35% profitability level with a bonus fund of 15% of the total wage fund (or approximately 60% of the fund of salaries for engineering-technical and office personnel).

These rough calculations make it clear that it is possible

to establish limits which serve as a guarantee against an excessive growth of bonus funds and assure that these funds will correspond to the reduction of production costs, i.e., to the growth of production efficiency.

Of course, these calculations are by no means ready-to-use norms, but serve only to illustrate a method for typologically characterizing enterprises. These characteristics must include the structure of outlays (with the given prices), the capital turnover rate, the structure of assets, etc. In planning, we make insufficient use of the method of long-term norms based on progressive practices, as we limit the comparability of enterprises to a great extent and do not bring out, through statistical analysis, the groups of enterprises which can be required by the state to achieve equal results as regards production efficiency. Meanwhile, special analysis — but without excessive differentiation of profitability norms — can produce, in our opinion, quite satisfactory results in working out long-term plan assignments.

It may seem, at first glance, that violations of production proportions can occur because of the enterprises' efforts to exceed their output plans. Actually, however, that is not the case. The enterprise can exceed its output plan only if it is given, under planning procedures, additional raw materials and, in some cases, additional equipment.

For this reason, the enterprise can and should strive to raise its profitability not only by expanding production, but also by taking better advantage of the intensive factors of production, i.e., by raising labor productivity, doing away with rejects, saving raw and other materials, cutting overhead expenses, etc. In this case the enterprise will not, as a rule, make additional demands upon the national economy which might violate production proportions.

Finally, let us assume that the above-plan growth of labor productivity releases a more or less considerable amount of manpower in the processing industries, the growth of which, however, is limited by raw material resources. In our conditions, this manpower can always be utilized in the mining industries, in construction, and in mechanized agriculture. And in this case there will be no disproportions.

All these factors constitute one of the guarantees that cost accounting in our industry will not be transformed into an instrument of spontaneous production regulation under the influence of the law of value.

How can all the types of material incentive be combined on the basis of the principle we have suggested? In our opinion, every enterprise must form (in accordance with the results of its operation for the quarter) an aggregate, single bonus fund depending on the fulfillment and overfulfillment of its planned long-term profitability limit. In every subsequent quarter, the enterprise must pay monthly bonuses to its managerial personnel out of this fund. However, the size of the bonus must depend not only on the degree to which the output plan is fulfilled, but also on the above-plan saving of means. A certain share of the fund may be spent, at the discretion of management and the trade-union organization, for one-time grants and on the cultural and everyday needs of the personnel. A portion of the fund should be kept in reserve till the annual report. And, finally, a certain share of this fund should be centralized on the account of the chief administration for replenishing the bonus funds of enterprises that are deprived of such a fund for reasons that are not connected with their efficiency, as well as for granting additional bonuses to enterprises that have done well in the all-Union socialist emulation.

Such a system would make material incentives clear and uniform. It would do away with the existing shortcomings in the progressive-bonus system of remuneration for the managerial staff. Today, for instance, if the enterprise has fulfilled its plan in all aspects, the engineering and technical personnel are given bonuses out of the wage fund. This means that these bonuses are included in production costs and thus make them higher. If, on the contrary, the enterprise worked inefficiently, then these bonuses are not paid out at all, though they are included in the production cost plan. This leads to a situation in which failure to fulfill the plan is a factor in cutting production costs for the item "wages." The cost of production becomes a distorted indicator of an enterprise's efficiency. The extent of the possible distortion is too great to be ignored: according to the 1954 data, bonuses to engineering and technical staff at the plants that have fulfilled their plans range from 0.6 to 1.5% of the total production costs, amounting at some plants to as much as 2.25% of these costs. This is a rather graphic example of how the disregard of economic categories leads to practical miscalculations. The sums that, in essence, are bonuses out of an enterprise's net income must not be included in the regular wage fund as a portion of the cost of the goods produced.

The procedure suggested here should lead not only to more precise planning, but to simpler planning as well. Planning in the chief administrations and ministries will be considerably more simple. There will not be any need to calculate cost and profitability limits on the scale of individual enterprises; attention should be focused on the proportionality of programs and their backing with materials and equipment. The plans for production costs and profits, drafted by the enterprises themselves, can be counted on to secure the branch assignments of the national economic plan and budget, and this is the only thing that will have to be checked. If, however, the branch plans of industry for a given year fail to correspond to the accumulation rate envisaged by the national economic plan, then some enterprises will have to be asked to raise their profitability level, with due consideration of the indices they have reached. But now this will not conflict with the interests of the enterprise's personnel, since such an increase in the plan assignment will not have an adverse effect upon the conditions determining the size of bonuses. On the contrary, higher profitability is stimulated by the progressively increasing share of profits allocated to the bonus fund.

It should be noted that simpler planning does not mean weaker planning. On the contrary, it strengthens planning. The more decentralized the planning is in its details, the more attention can be concentrated on the key items of the plan. Use of the principle of material incentive and cost accounting for improving planning on the basis of local initiative will produce better results than the attempts to calculate all the indices of the enterprise's operation at the center. The former method represents, from the scientific point of view, a higher level of planning, not a lower level.

All this, of course, is but a rough outline of a solution to the problem. There are probably much better variants. As always in such cases, it is essential to discuss and test the new methods of combining planning with material incentive, even if only at a limited number of enterprises. If it has been possible to test new methods of crediting turnover and methods of rewarding reductions in production costs for a number of years at certain enterprises, then it is certainly worth verifying the new methods in this decisive area. It is decisive because it involves the finding of ways to strengthen cost accounting and, at the same time, to disclose the reserves of production and simplify the administrative machinery.

Footnotes

1) Ia. Kronrod, <u>Osnovy khoziaistvennogo rascheta</u>, Gos-finizdat, 1952, p. 15.

2) Kronrod himself mentions some cases (p. 215 of his book) of enterprises operating at a high profitability level while failing to fulfill their plan assignments for production costs.

Kommunist, 1956, No. 10

Planning Industrial Production and Material Stimuli for Its Development

E. G. Liberman

The 20th Congress of the CPSU posed a number of big and important tasks in the concrete economics of industry. Our economic science has a special role to play in the accomplishment of these tasks. It must seriously tackle problems of simplifying the system of production management, improving planning, and strengthening the personal material interest of employees in the results of their work. All these problems are interrelated and cannot be considered separately from the main task, that of raising the productivity of labor on the basis of technical progress. The machine-building ministries are now drawing up regulations on bonus payments for introducing new technology on the basis of the approved standard regulations. This will undoubtedly advance equipment and technology, and facilitate implementation of the measures planned by the State Committee on New Technology and the machine-building ministries.

However, this does not release our economic science from the need to examine comprehensively the question of how best to employ the principle of personal material interest for stimulating the further rapid growth of labor productivity and reducing production costs. Material stimuli will help constantly to reveal and press into service the inexhaustible reserves of socialist production. Technology is improved not only by applying more or less big, centrally planned measures, but also by the day-to-day, painstaking and persistent work of enterprises in mastering existing technology, modernizing equipment, and carrying out small-scale mechanization and automation of work processes. There are also vast possibilities for the growth of labor productivity and the reduction of production costs in a steady improvement of the organization of production. All this calls for examining anew the problem of material stimuli in industry as a problem of improving planning methods in industry.

The principle of personal material incentive largely operates where we have introduced piecework remuneration for workers (though here there are many shortcomings which should be removed as quickly as possible). As regards engineers, technicians, and administrative personnel in all branches of management, the situation with regard to incentives is still less satisfactory. Meanwhile, the growth of labor productivity depends in large measure on the quality of the management of shops and enterprises, on the material interest of the personnel.

The socialist mode of production, as compared with the capitalist one, provides immense advantages for technical progress and constant growth of production. This has been shown by our rapid advance to the summits of science and the extensive application of scientific achievements in the socialist economy. Labor productivity in our country is now rising by 6 to 8% a year, that is, at rates which capitalism never achieved even in its most successful period. Moreover, production in our country goes up chiefly owing to the steady rise of labor productivity. Per capita output is growing continually. Thus, per capita industrial output in the USSR increased 19.4 times from 1913 to 1955, while in the USA it grew only 2.3 times. This is primarily a manifestation of the advantages of economic planning and the new socialist attitude toward work.

But if we examine the everyday facts of production, we find that the potential of our socialist economy is far from being fully utilized. True, the volume of production is increasing constantly, but qualitative changes in the methods of production are being made rather slowly, and mainly by investments in building new enterprises or reconstructing existing ones. After new technology has been introduced, a lull often sets in at the given factory or plant and there is a tendency to maintain production on the achieved qualitative level for a long time, which is economically unjustified. Output at such enterprises grows chiefly by employing easily mobilized reserves and partially eliminating obvious production losses. Therefore, production costs are lowered mainly through a relative reduction in overhead resulting from an increase in the volume of output and without effecting sufficient organizational and technical measures.

To confirm what we have said, let us cite some data on

production costs, their composition, and itemized reductions at Kharkov mass-production machine-building factories for 1953 and 1954. Production costs of comparable output consisted there (at constant wholesale prices) of the following expenditures: basic materials — 53%, direct wages — 11%, overhead — 36%. Production costs appear to have gone down rather substantially — 10% a year, including shop and general factory expenditures by 6.3%, wages by 1.1%, basic materials by 2.6%. These figures indicate a disproportionately small reduction in the consumption of materials, which was due to a lag in the technology of preparatory operations and to a low metal utilization factor. The "secret" of the reduction in production costs is simple: the volume of output increased by 18 to 21% a year, and this produced a high index of cost reduction chiefly through the lessening of overhead.

Quantitatively our output is rising at rapid rates, which certainly enhances the effectiveness of production. But this is not enough; there must be constant improvement of the methods of production. Many enterprises effect qualitative changes in production rather slowly, with difficulty; they are often resisted by personnel who do not wish to abandon the old ways of doing things. Such conservatives were sharply criticized at the All-Union Conference of Industrial Personnel, the July 1955 Plenary Meeting of the CPSU Central Committee, and the 20th Party Congress.

Economists constantly stress the importance of technical progress, of the specialization of enterprises, shops, and sections. But economic science does not adequately bring out the reasons for the gap between our vast potential for raising the effectiveness of production and the realization of this potential in practice. Indeed, we cannot attribute everything to the resistance of conservative people: in the first place, industrial personnel in many cases provide models of mature engineering solutions and selfless work; second, if inertness and conservatism are encountered too often, the reason should be sought in shortcomings in production planning and management which create the soil for such an approach.

Planned management of industrial enterprises and the consequent relationships between the people taking part in production cannot but undergo essential changes with the development of the country's productive forces. Such changes have taken place throughout the period of the economic development of the USSR. Taking into account the conditions and requirements of economic development, the Communist Party's directives have always envisaged improvements in the forms and methods of the planned management of production.

On the basis of party directives, a number of major measures in this field have been carried out in recent years. Long-term and current planning has been placed in separate agencies; democratic centralism has been strengthened in economic administration; many industries and enterprises have been transferred from all-Union administration to that of the respective republics; the powers of the ministries, chief administrations, and factory managers have been expanded.

All these changes reflect, first, the tremendous expansion of production, given which the excessive centralization of management hinders the development of industry; second, the fact that the local managerial personnel are now much more mature, conscientious and competent, which makes it easier to draw on local experience and initiative and local peculiarities for improving production in the country as a whole.

This creates the conditions for further improving the planned management of the economy. The existing planning procedures still lag, in many respects, behind the requirements of the growing socialist economy. The gigantic tasks of quantitatively and qualitatively developing industrial production cannot be accomplished by the methods of planning and material stimulation that were effective for a certain stage of economic development, but can hinder progress in the new stage.

"The whole point," Lenin said, "is not to rest content with the skill we have acquired by our previous experience, but without fail to go on, to strive for something bigger, to proceed from simple tasks to more difficult ones. Otherwise, no progress is possible in socialist construction" (Soch. [Collected Works], Vol. 28, p. 172).

Specifically, the major economic task of unleashing the initiative of enterprises in the matter of utilizing their latent production reserves cannot be accomplished by merely broadening the administrative powers of their managers. For no matter how broadly these powers are formulated on paper, they cannot be applied in practice unless they are backed by appropriate economic measures. Here we have in mind, above all, improvement of the planning of enterprise operations and the allocation to enterprises of sufficient funds for material incentive. This is in full accord with the decisions of the 20th CPSU Congress, which called for simplifying production management, improving planning procedures, and heightening the material interest of industrial personnel in order to give full scope to continual improvement of production methods.

This requires a profound study and analysis of concrete forms of planning the economic indices of our enterprises.

***

Everyone now agrees that the targets set for enterprises from the center should be limited in number and not be too detailed. From this standpoint let us examine the methods of planning economic indices.

As regards quantitative targets, i.e., the volume of output in specified items, these targets should unquestionably be communicated to enterprises annually, with a breakdown by quarters. This is an essential condition for the growth of production and for ensuring its proportionality, for maintaining the necessary rates and trends of the country's economic development. Specified targets for many lines of production or groups of goods may be set from the center by the ministries only in consolidated units; the details and time limits for delivery should be the subject of direct contractual agreements between suppliers and customers.

It is essential that the nomenclature of output be renovated by the mastery of more productive instruments of labor, more economical objects of labor, and products of higher quality. This will be facilitated by the system, now being adopted, of awarding bonuses to machine-building industry personnel for introducing new technology. But the change-over to the output of new goods must improve, rather than worsen, the performance indices of enterprises, particularly their profitability.

The trouble is that the generally applied method of fixing wholesale prices by adding minimum profit (3 to 4% of the production cost of a machine) not only does not stimulate the introduction of new machines; it retards the process. Factories usually attain a much higher profitability in putting out machines that have been in production for a long time. Assuming that in the first year after the establishment of new wholesale prices the average profitability was taken to be 3%, then profitability for mastered comparable output will rise approximately as shown in the table on page 25.

This table needs some explanation. Wholesale prices in machine building are fixed by adding a profit of 3% to the cost of production. Hence, if we take the price as 100%, it will consist in the first year of 2.9% profit and 97.1% production cost,

Year after establishment of wholesale prices	Production cost (% of price)	Profit (% of price)	Profitability (% of production cost)
1st	97.1	2.9	3.0
2nd	94.2	5.8	6.16
3rd	91.2	8.8	9.65
4th	88.4	11.6	13.1
5th	85.7	14.3	16.6

since $\frac{2.9}{97.1} \times 100$ comprises 3% profitability. Further, if in the first year the cost of production comes down only by 3%, then in the second year it will be lowered by $97.1 \times \frac{3}{100}$, or by 2.9% of the price. Thus, in the second year the cost of production will comprise $97.1 - 2.9 = 94.2\%$ of the price, and profit will be $100 - 94.2 = 5.8\%$ of the price. Relative to the cost of production, this profit will equal $\frac{5.8}{94.2} \times 100 \approx 6.16\%$ profitability, and so on. These calculations show how profitability in the output of mastered goods rises if the cost of production declines by only 3%.

Even at such minimal rates of cost reduction, there is no economic sense for an enterprise to shift to the output of new machines; indeed, when it puts out long-mastered goods its profitability in any year of the five-year period will be much higher than if it manufactures new machines. For example, in the third year of the operation of the wholesale prices, an enterprise putting out old models of machines will have about a 10% profit, whereas for new machines the wholesale price will be fixed on the basis of a profit of 3% of the planned cost of production, although in this case it is harder to keep costs within the prescribed limits. We take absolutely no account of the technical risks, additional outlays, and delays in fulfilling output plans that are encountered by enterprises in shifting to the production of new, generally more complex machines. Consequently, enterprises mastering the output of new equipment are placed in worse conditions than those producing old models.

To gain better conditions for the output of new models, many factories seek to obtain excessive plan production costs.

They avail themselves of the fact that on new items no targets are set for reducing costs in comparison with the previous year. Moreover, in the next year enterprises easily attain, on new items, a considerable reduction of production costs relative to the overstated base cost of production. What does all this show? It shows that the methods of planning costs separately for comparable and incomparable goods, proceeding from the report "base" of each enterprise, are unsatisfactory and unreliable.

It is usually argued that a decline in profitability connected with a shift to the output of new machines is envisaged in the annual plan and that the enterprise suffers no visible losses from this decline. But the fact is that an enterprise's incentive fund is formed in accordance with certain norms (in machine building, for example, 4% of planned profit and 40% of above-plan profit). This means that if, as a result of a high proportion of new machines, an enterprise's profit is reduced by the plan, the enterprise suffers losses — and substantial ones at that.

Our existing forms of material incentives for enterprises are inadequately dovetailed with the need to stimulate the advancement of production. The latter is also not facilitated by our system of price formation. Instead of providing enterprises with stimuli for putting out new machines, we "fine" them for it in concealed ways. Indeed, the mastery of a new model nullifies all that the enterprise has achieved over a number of years and drops it back to the level of the first year of operation of the new wholesale prices. The "time factor" is ignored here; account is not taken of the fact that enterprises raise their labor productivity from year to year, which finds expression in lower production costs and higher profitability with the same wholesale prices. It is evident that new solutions must be sought here too, solutions based on a deeper analysis of the indices of the effectiveness of production in their interrelation.

This will become still clearer if we pass from quantitative indices to qualitative ones — to labor productivity, production costs, and the profitability of enterprises. There must be a common approach to planning these indices because, in different forms, they all express the same thing: the effectiveness of the expenditures of labor (past and living).

We do not have such a common approach. Labor productivity targets (in the form of output per worker) are given to an enterprise as an economic directive, as a consolidated index depending on the enterprise's level of technology and organization of production. On the basis of this plan target the enterprise, in its technical, production and financial plan and in the course of its fulfillment, adopts such a technological process and such output norms as to attain the labor productivity target, irrespective of revisions in the assortment of its output.

There is an entirely different approach to planning the cost of production and profitability of an enterprise. Item-by-item estimates of expenditures on production are given to enterprises by the central agencies. In no lesser detail is calculated the cost of the whole commodity output, reduction in the cost of comparable goods (in percentage and in absolute figures), and sometimes even calculations on separate articles. Still greater detail is provided in the centralized planning of profit and profitability, since here the attempt is made to proceed from the difference between the value of the output sold in wholesale prices and its production costs for each single item.

Such detailed centralized planning of production costs and profit presupposes a precisely fixed assortment of output for the next year, set technological processes, and norms of outlays of labor, materials, tools, and power. Owing to this the directive targets for costs of production are subordinated to numerous calculations submitted by the enterprises and are checked on the basis of the report data of each enterprise with certain (usually "volitional") revisions. Instead of singling out that which is common and regular (and this should be the guiding principle for a whole group of similar enterprises), centralized planning is at times bound by the practices prevailing at each enterprise.

Hence, the economic directive on these qualitative indices ceases to be the guiding principle. It is subordinated in one degree or another to the technology and organization of production existing at the various enterprises. Yet numerous facts show that identical goods can be produced by quite different technological processes even at similar enterprises, in connection with which they can cost much more or much less to produce.

This question arises: is it necessary to determine from the center, proceeding from such detailed calculations, the cost of production and the profit for each enterprise in industries producing a wide range of goods? Seemingly "accurate," such targets usually turn out to be fundamentally inaccurate. And this is largely responsible for the plans of enterprises being poorly coordinated and divergent, since it is technically impossible to draw up a well coordinated plan when some indices are given in detail and others in consolidated units.

It should also be noted that the production cost is generally not expressed in our plans by a single index. It is defined as an estimate of expenditures on production, or as a calculation of units of commodity output, or as a percentage of reduction of the cost of comparable goods. For centralized plans, however, it is necessary to establish a single consolidated target, an economically substantiated directive on production outlays in terms of value.

Concrete economics as a science has not sought such indices. Economists do not see a contradiction in the fact that such a major index as labor productivity may be specified for enterprises as a ratio between the volume of gross output in wholesale (constant) prices and outlays of labor time, whereas production costs cannot, for some reason, be projected in such a consolidated form as a ratio between the volume of output in the same wholesale prices and all money expenditures on the production of this output.

One possible variant of a qualitative standard may be the index of profitability of production. This would furnish a uniform, generalized expression of production costs as a magnitude of money expenditures per unit of output. This magnitude is fully determined by the average profitability of production. Production costs and profitability are interdependent and mutually determining magnitudes.

This can be illustrated by the following example. If we take the value of total output in wholesale prices at 1,100,000 rubles and production costs at 1,000,000 rubles, the average production costs (per ruble) will be:

$$\frac{1,000,000 \text{ rubles}}{1,100,000 \text{ rubles}} \approx 90.9 \text{ kopecks}.$$

The converse magnitude multiplied by 100 produces:

$$\frac{1,100,000 \text{ rubles}}{1,000,000 \text{ rubles}} \times 100 = 110\%.$$

This is nothing but profitability plus 100%, for profitability equals:

$$\frac{1,100,000 \text{ rubles} - 1,000,000 \text{ rubles}}{1,000,000 \text{ rubles}} \times 100 = 10\%.$$

It is assumed here that the wholesale constant prices and the operating wholesale prices are identical, as they should be, while the same wholesale price list is in force.

Since the assortment of goods has to be taken into account,

it seems impossible to determine production costs for planning purposes by such a consolidated factor as average profitability (or, what is the same thing, average cost per unit of output calculated in wholesale prices). But we are simply accustomed to planning costs of production on the basis of expenditures on each article. Therefore, every change in assortment alters the average profitability (or average cost per unit of output) both in the plan and in fact. Actually, however, there is no need to envisage changes in planned profitability in connection with shifts in assortment, if the type or character of the enterprise has not altered radically, if it has not been substantially reconstructed. We should note, incidentally, that the assortment sharply fluctuates at our enterprises because we do not as yet have sufficient specialization. In conditions of broad specialization, an enterprise, even if its output assortment is changed, can and should maintain its average planned profitability by reducing the cost of producing particular goods. And it will do so if it has a real material incentive to promote the profitability of production. At present, if a change in assortment raises actual profitability over the planned figure, an enterprise prefers to keep quiet about it. On the other hand, if profitability falls below the assigned figure, the enterprise shouts for a revision of the plans. However, this is usually of no avail since the budget has already been adopted and no one will venture to lower the profit figure set by the plan.

The whole confusion of mutual claims and calculations will disappear if we take as a guide the average profitability of the given type of enterprises, irrespective of any assortment changes. At the same time, in consumer goods industries, higher profitability from the output of goods in great demand and short supply should be stimulated through a system of prices.

It would be interesting to know the real dynamics of the average cost of a unit of output, to what extent this magnitude depends on the types of enterprises, and in what measure assortment changes influence it. Our economics literature has not yet given an answer to this question.

Meanwhile, as investigations by the author show, production expenditures per ruble of gross output in wholesale prices (which we shall designate here as the average cost of a unit of output) depend in large measure on the type of production.

This is evident, for example, from the data for fifteen Kharkov machine-building factories (see table on page 30).

The data in this table do not at all imply that all factories of the same type have absolutely equal expenditures per ruble of output; there are fluctuations here. Nevertheless, their grouping by types of production reveals the influence of each type on the average cost per unit of the physical volume of output. This reduction in the average production cost is tantamount to a rise in profitability and is naturally connected with the growth of labor productivity.

But even if cost per unit of output did fluctuate sharply because of assortment changes, would that mean that profitability is a secondary or derivative factor while the cost of individual articles is the primary and basic factor? Would it not be more correct, in setting state centralized targets, to take the generalized index of profitability as the starting point from which the cost of individual products specified in the technical, production and financial plan of an enterprise would be reckoned, taking into account the concrete output assortment? That would be more consistent, for in this case we would be planning labor productivity and production costs by the same method. If we can plan labor productivity as a generalized economic directive for the expenditure of living labor per unit of output, we should in the same manner plan profitability as a generalized expression of production costs or money expenditures on the same unit of output.

We by no means wish to say that labor productivity is planned in our country with ideal precision. It is well known that this index too is affected by the imperfection of our wholesale prices as measures of the physical volume of production. Hence, price formation must be fundamentally improved; this

Average Cost of a Unit of Output

(per ruble of gross output in 1952 wholesale factories)

	1952	1953	1954
	(in kopecks)		
Single-unit or small-batch production enterprises (4 factories)	107	95	91
Batch-production enterprises (5 factories)	93	87	84
Mass-production enterprises (6 factories)	91	83	79

will raise the quality of the planning of labor productivity and profitability of production. Improvement of price formation is closely linked with reassessment of operating fixed assets in accordance with their real (restoration) value. This reassessment is long overdue. It is an imperative requisite for a correct calculation of depreciation funds and, hence, also production costs. The question of price formation is quite complex and is beyond the scope of this article. Nevertheless, our economic science must be required to devise an objective basis for substantiating prices and their deviation from value.

In improving planning procedures, it is first of all necessary to find a simple, generalized qualitative norm of expenditures on production. In the Soviet economy, where the law of value operates, profitability of production, as we have noted above, may serve as that qualitative norm. Of course, one can and should seek other generalized qualitative norms. To illustrate that it is possible in principle to solve this problem, we shall attempt here to substantiate the profitability norm. If we accept this qualitative norm, it will be sufficient to plan for enterprises from the center only one index, profitability, instead of several indices of production costs. For enterprises producing similar goods in approximately similar conditions, one and the same level of profitability will be planned. The profitability index will already take into account the money expenditure on the production and sale of the goods. The growth of labor productivity, average wages, the limits of administrative expenses, etc., will also be planned centrally.

Then the profitability target (i.e., average production cost target) would be the basic factor, while the technology of production, norms of output and of expenditure of materials, and, hence, the cost of individual articles, would be derivative factors. They should be established by the enterprise itself in its plan, and in such a manner as to attain the state targets for the volume of output as well as the qualitative norms for labor productivity and profitability. This would mean that it is not the cost of production of individual articles which would determine profitability but, on the contrary, that cost would be determined by fulfillment of the state profitability targets, as they express the advanced experience of similar enterprises.

Such a system would improve the performance of enterprises. It would also relieve the planning bodies of unnecessary and at times even harmful detailing of calculations and would elevate planning to a higher qualitative plane.

From the center each enterprise should be given an <u>aim</u> in the form of an output plan specifying definite progressive qualitative norms of labor expenditures (depending primarily on the type of enterprise and only in some cases on its assortment of output). And it should be given leeway in time and material incentive for the earliest possible attainment of the progressive norms.

How is such leeway to be ensured? Obviously, the qualitative norms should operate for a long time, say, as long as the same wholesale prices are in force; these prices should, as a rule, also not be changed more often than once in five years.

The need to go over to planning for long periods is well realized. Plans for long periods should, however, be not only guiding but also operative instruments. At the 12th Moscow Regional Party Conference, the director of the Kauchuk Factory, I. M. Manvelov, justly said: "Reduction in production costs is now planned only in comparison with the preceding year. As a result, poorly functioning enterprises are put in a more favorable position than those which carry out their plans." In this connection, Manvelov suggested that production costs for the year 1955 be taken as the basic index and serve until the end of the five-year period. This is a recognition that the norm should be a long-term one, the only difference being that in the form of production costs the norm is suitable for comparable goods, while in the form of profitability it embraces the whole volume of commodity output irrespective of its comparability.

***

The 20th CPSU Congress recognized that long-term plans should be drawn up for each enterprise. Labor productivity is actually already being planned for enterprises for a five-year period. That being so, profitability too can be planned for a five-year period. The growth of profitability should be planned like the growth of labor productivity — by years for the whole five-year period.

Let us examine profitability as an economic indicator in index form, i.e., as a relationship between the volume of output in wholesale prices and the cost of production of the same volume of output. Then profitability can be represented as the product of a number of coefficients (factors).

That it is correct to break up profitability into factors is proved by the fact that when the like-named magnitudes are reduced, the equation becomes an identity.

$$\frac{\text{Output in wholesale prices}}{\text{Cost of production}} = \frac{\text{Total wage fund}}{\text{Cost of production}} \times \frac{\text{Workers' wages}}{\text{Total wage fund}} \times \frac{\text{Number of workers}}{\text{Workers' wages}} \times \frac{\text{Output in wholesale prices}}{\text{Number of workers}}$$

Hence, profitability depends: 1) on the share of total wages and salaries in production costs; 2) on economical management, i.e., on the share of workers' wages in the total wage fund; 3) on average workers' wages, or rather on their converse magnitude; and 4) on the productivity of labor.

The data for the machine-building enterprises show that the first coefficient, the share of wages and salaries in production costs, is fairly stable for similar enterprises, notwithstanding changes in the assortment of their output. Thus, at the Kharkov Tractor Plant it was 0.31 in 1952, 0.317 in 1953, and 0.311 in 1954. For the whole group of mass-production factories, the share of wages and salaries in production costs changed by years as follows: 0.295 in 1952, 0.295 in 1953, 0.282 in 1954. The coefficients were higher in the batch-production factories: 0.328, 0.315, and 0.301. They were still higher at the single-unit and small-batch production enterprises: 0.371, 0.422, 0.388. The share of total wages and salaries in production costs is also a characteristic feature for individual branches of the machine-building industry. Thus, in general batch-production machine building it was 28 to 30%; in machine tool manufacturing — 35 to 40%; in precision instruments production — 50 to 60%, etc.

Another factor of profitability, the share of workers' wages in the total wage fund, shows how economical the management is. The higher this share, the less, relatively, is spent on remunerating the technical, engineering and office staff, and the higher, consequently, is the profitability of the enterprise.

The table on page 34 contains numerical expressions of this coefficient for 1952-1954 at the Kharkov machine-building factories.

These data testify quite convincingly to the advantages of specialized enterprises. At the same time, it should be noted that advanced enterprises of each type have more economical indices than the other enterprises of the given type. At the fuel

Type	1952	1953	1954
Mass production	0.767	0.769	0.767
Batch production	0.751	0.756	0.752
Single-unit and small-batch production	0.677	0.675	0.674

equipment factory, workers' wages comprise 82% of the total wage fund (the average for the mass-production enterprises is 77%). Among the batch-production enterprises, the Svet Shakh-tera Plant has a coefficient of 0.80 (against this group's aver-age of 0.75). In the small-batch group, the Gioroprivod Factory has attained a coefficient of 0.75 (the group's average being 0.68). It follows that it is quite possible to envisage long-term targets for raising the coefficient of economical management by drawing on the experience of the advanced factories of the given group — and that irrespective of output assortment changes.

The same is also true of the third profitability factor — average workers' wages. This magnitude should be determined on the basis of general economic considerations and with ap-proximately equal rates for a wide range of similar enter-prises. Otherwise we shall never eliminate the differences in pay for equal work.

Now we come to the fourth, and decisive, profitability factor — labor productivity, i.e., output per worker. Labor productivity must always grow faster than average wages, for otherwise society will not be able to create that part of the national income which is necessary for expanded reproduction and for meeting social needs. That is precisely why labor productivity is the decisive factor of profitability.

It now becomes clear that long-term or long-operating norms of profitability can be determined with sufficient accu-racy on the basis of typical characteristics of enterprises by branches of industry. These long-term norms give an enterprise a clear perspective and make it strive for economy, for genuine cost accounting. It will no longer fear that a reduction in production costs this year will evoke a revision of its qualita-tive target in the plan for next year.

How then is this norm to be calculated by years of a five-year period? Let us take mass-production enterprises with a characteristic feature of total wages comprising 30%.

of production costs (in other words, with a coefficient of 0.3). According to the law established by Marx, this percentage should go down as labor productivity rises. In many industries, however, and in machine building in particular, it is necessary to proceed from the need to maintain it on a constant level for a number of years. This ensures an even reduction of production costs from outlays of both living and past labor. We have considerable possibilities for saving metal — by reducing the weight of machines, applying progressive technologies in the preparatory processes, etc. These possibilities are by no means smaller than those connected with lowering the expenditure of living labor. The demand that in machine building at this stage the share of the total wages in the cost of production should be maintained at a constant level is a progressive one; it leads to a higher profitability of production.

Suppose that this share remains stable; then the coefficient of change by years of the planned five-year period will be 1.0. (For other industries or groups of enterprises, of course, this coefficient may be expected to change in the coming five-year period in dependence on the relationship between the possibilities for reducing labor requirements and materials requirements per unit of output.)

Let the share of workers' wages in the total wage fund equal 0.75 (which, as we have seen, corresponds to the actual relationship). The state calls for greater economy in management and, proceeding from the achievements of the best factories, plans this coefficient to rise to 0.80. This means that in five years it will go up by 7% $\left(\frac{0.80}{0.75} \times 100 \approx 1.07\right)$. The average annual change coefficient will amount to 1.004; i.e., there will be an annual rise of 0.4% ($\sqrt[5]{1.07} \approx 1.004$). This change does not depend on alterations in the output assortment.

Suppose, further, that general economic considerations, not the output assortment, dictate a 14% nominal rise in average wages over the five-year period. A change of the converse magnitude will equal $\frac{1}{1.14} \approx 0.88$, and the average annual change coefficient will be 0.975 (since $\sqrt[5]{0.88} \approx 0.975$). Lastly, let the planned rise in labor productivity in the five-year period be 40% (again, regardless of changes in the assortment of similar goods). This means that the average annual change will equal $\sqrt[5]{1.40} \approx 1.06$, or an increase of 6% a year.

Now we can determine the rate of growth of profitability by

years of the five-year period as a coefficient:
$$C = 1 \times 1.004 \times 0.975 \times 1.06 = 1.038,$$
which means an annual growth of profitability by 3.8%.

Thus, two targets, an annual increase in labor productivity by 6% and an annual rise of profitability by 3.8%, will be the basic long-term qualitative norms.

If, in the year taken as a basis, the leading enterprises had an average profitability of 1% (without balance losses) expressed in wholesale prices, then profitability should go up at similar enterprises in the years of the planned five-year period as follows:

1st year — $1.01 \quad \times 1.038 \approx 1.0484$ (roundly 5%)
2nd year — $1.0484 \times 1.038 \approx 1.088$ (roundly 9%)
3rd year — $1.088 \quad \times 1.038 \approx 1.129$ (roundly 13%)
4th year — $1.129 \quad \times 1.038 \approx 1.172$ (roundly 17%)
5th year — $1.172 \quad \times 1.038 \approx 1.217$ (roundly 27%)

This, of course, is only a tentative pattern of possible calculations. But it shows how a generalized long-term norm for a qualitative improvement of production can be substantiated on the basis of a comparative analysis and deduced laws of economic development.

At first glance it may seem that we have here the old subjective, "volitional" approach to determining long-term qualitative norms, since we proceed from a base profitability. This may prompt an enterprise deliberately to lower its "base." It should be noted, however, that instead of reports by each individual enterprise, our base is to be made up of indices for a group of similar enterprises, with an orientation toward the stable characteristics of the leading and best enterprises. This will make for objective planning, since objective characteristics can be revealed only by studying an aggregate of enterprises, and this precludes chance and arbitrariness in establishing norms.

Some may doubt that the planning bodies can accurately establish long-term norms, and then enterprises would be placed in unequal positions — some in better, others in worse positions. We believe that these fears are groundless. In the first place, even with the existing planning methods, enterprises are placed in quite different positions. Second, the magnitudes of the incentive fund and of individual bonuses are limited in advance by a certain maximum. Third, and this is the most important point, all enterprises will strive to attain a progressive norm of profitability, which will make for a general growth of production. The necessary degree of accuracy in planning is achieved by correctly combining consolidated centralized targets with detailed, technically substantiated planning at enterprises. Should essential inaccuracies in the norm for a given enterprise be revealed, then the appropriate bodies (outside the system of the given ministry) will be able, by way of economic arbitration, to decide the question of revising the norm before its term of operation expires.

Some readers may arrive at the idea that it might be better to seek an objective qualitative norm in the form of output per unit of production assets or even in the form of a rate of profit in percentage of the value of these assets. We do not preclude the search for diverse objective gauges of production effectiveness. Our concrete calculations serve only to illustrate possible solutions for this complex problem. Let us only note that output per unit of assets will inevitably reflect inaccuracies not only in evaluating output at wholesale prices, but also in evaluating the fixed assets. As regards profit in percentages of the value of applied assets, this (leaving aside the fundamental aspect of the matter) would be influenced in still greater measure by price formation factors and inaccuracies in evaluating the assets.

On the other hand, by grouping enterprises according to type, primarily according to the share of wages and salaries in the cost of output, we, in essence, reflect in actual prices the organic structure of their assets. This share declines with the growth of the organic structure of production assets. Hence, our profitability norms, established for types of enterprises, already contain the objective factor of a stable relationship between the expenditures of living and past labor, which is peculiar to the given type of production and which predetermines the extent to which the growth of output will be expressed through the growth of profitability.

Determination of long-term norms of profitability is one of the possible ways of fundamentally improving the system of material incentives for enterprises. At the present time, the size of the incentives depends on the fulfillment of current plans. This system undoubtedly stimulates efforts to fulfill the plans. But it also produces serious negative consequences: disclosure and use of reserves in the given year worsen the incentive conditions in the next year, and this retards the growth of production, the introduction of new technology, and the reduction of production costs. Today enterprises do not have a material interest in taking on demanding plans. They are interested in the opposite. Plan assignments are often distributed among factories without proper substantiation, in accordance with the "resistance" of the given factory. Enterprises do not have a material interest in good performance, since this results in the transferring to them of assignments from enterprises which are functioning poorly; and the end result is that good enterprises become "bad" ones.

Advanced enterprises have their plans increased late in the year while the plans of lagging enterprises are reduced. Thus, the Kama Pulp-and-Paper Combine had its output plan for 1955 altered on December 30 of that year, and its final revised financial plan was approved in January 1956. Both plans were so "rectified" that the combine was converted from an enterprise which had fulfilled its accumulation target and had an above-plan profit of several million rubles into one that had fallen behind on these indices: its above-plan output dropped by 7,000,000 rubles. And yet its original profit targets were probably calculated with the utmost "accuracy" for its entire output assortment. Would it not be simpler to set the combine a long-term norm of profitability for five years and not alter this norm for the purpose of showing an even performance by the ministry as a whole?

The fact that the result of any innovation is immediately included in the next plan leads to an absurdity: the greater the effort made by an enterprise to introduce new methods, the harder it becomes for it to get incentive funds in the future. Given these conditions, can one expect the personnel readily to take technical risks, to improve constantly the equipment and technology of production? It is easier for them to fulfill the plan by turning out mastered goods with the technology they are accustomed to, though it may be outdated. As we see, if the size of incentives depends only on the fulfillment of current plans, this often retards the progress of production.

The situation would alter essentially if the size of incentives depended on the degree of fulfillment of progressive qualitative norms that have been set for a number of years. Adoption of stable norms is obviously the surest way to enhance material interest so that it will operate not only when the current plan is being fulfilled, but will also serve to stimulate an enterprise to take on sufficiently demanding plans and apply the most effective production methods for their fulfillment. It is no accident that in order to stimulate collective-farm production, the Communist Party has adopted the per-hectare principle of calculating obligatory deliveries: per-hectare calculation is in fact the principle of stable norms applied to collective-farm production.

It goes without saying that industrial enterprises should gain the <u>right</u> to material reward only if they fulfill the current plans (and contracts) with regard to quality and volume of output in assortment. But the <u>size</u> of the reward must depend entirely on the degree of fulfillment of the long-term qualitative norms for years of the five-year period and types of enterprises.

Under such a system the enterprises will, first, have a material interest in taking on bigger rather than understated output assignments, for higher output is the surest way to increase profitability rapidly. In machine building, for instance, every 10% increase in the volume of output by itself yields a reduction in costs of at least 3%, and we have already seen how rapidly this raises profitability by years of the five-year period. Under our conditions, an enterprise has a sure way of sharply increasing production — by taking on a high output plan, as only the plan ensures it the required supply of raw materials, fuel, power, and other materials.

Second, enterprises will not fear to reveal and use their reserves, as this will not alter the fixed terms of material incentive for the future. Good performance this year will no longer be the cause of "bad" performance next year.

Third, an enterprise will do its best to effect technical and organizational innovations as extensively and quickly as possible, since they will unfailingly help to accelerate the growth of profitability and, hence, the growth of the incentive fund. The technical risk and outlays on mastering new processes will pay off regardless of when they bear fruit.

And, finally, this system will considerably simplify planning in the ministries; it will relieve them of unnecessary regulation and detailing of factory plans, entrusting this matter to the enterprises on the basis of their own material interest. This fully conforms to the spirit and meaning of the decisions of the 20th CPSU Congress and accords with the increased initiative and technical maturity of the local managerial personnel.

The principle of long-term norms will not only not weaken annual planning, it will strengthen it. Enterprises will continue to get assignments every year for their volume and assortment of output. But having long-term qualitative norms, they will draw up their own technical, production and financial plans, incorporating all quantitative and qualitative indices. These plans will also include estimates of expenditures on production and, based on these estimates, financial indices for circulating assets and size of profits, i.e., a balance sheet of income and expenditures. When an enterprise or its shops are to be reconstructed, an estimate will be drawn up of funds required from the state budget if the enterprise's own depreciation fund and profit are insufficient to finance the approved projects. In the same manner, enterprises will draw up requests for financial outlays to master new lines of output in conformity with adopted plans for introducing new types of goods. Central planning bodies will only have to check in what measure the consolidated plans for branches of production conform to the requirements of the budget and the national economic plan.

With the introduction of long-term norms, enterprises will be directly interested in having labor productivity rise, production costs decline, and profitability go up. Under these conditions, fulfillment of the qualitative indices envisaged in the annual national economic plans for branches of industry will be ensured by the technical, production and financial plans of enterprises. Even if, by way of control, it should be necessary to revise some annual indices in the technical, production and financial plans of individual enterprises, this would not alter the basic norms for rewarding enterprises and therefore would not be resisted by them at all. Hence, the drafting of annual national economic plans will be simplified, based on local initiative and material interest, and aimed at utilizing production reserves.

For greater clarity let us resort to the following graphic illustration. Each year an enterprise ascends, step by step, the ladder of progress. At present, the higher it goes, the more is demanded of it in the next year. And it does not know in advance what height it should reach. If it happens to fall off some step, this, after some unpleasantness, may prove to be "advantageous," as the enterprise will now begin to get assignments to lower steps. Or it will itself try to move more slowly, stepping each time on the rather low step it has succeeded in securing through "bargaining." On the other hand, if the plan provides in advance for the whole ladder, consisting of five steps, to be ascended in five years, though the steps grow progressively higher each year, then we can firmly tell the enterprise: should you go up faster, say, reach the fourth step in the third year, you will get a substantial reward for it. Then the enterprise will apply all the technical and organizational means at its disposal to achieve the fastest possible rise in labor productivity and reduction of production costs: it will be materially interested in such advancement. The enterprise will have before it an economic perspective which is now circumscribed by the current plans.

Qualitative norms for shops should be set for a period of not less than a year so that the effect of organizational and technical measures will not be calculated in the plan of each successive quarter and thereby lessen the stimulus for technical improvements by the shop personnel.

Later on it would be appropriate to have a single incentive fund so that there will not be diverse sources and terms of bonus payment. All bonuses (regular, monthly, and occasional for developing and introducing new techniques) should be paid to innovators of production — workers, technical and engineering personnel — from a single fund. It is advisable that payments into the incentive fund be made in accordance with the results of each quarter so that the fund should be the source for paying all bonuses in the subsequent quarter. The bulk of the fund should be used to pay bonuses for the introduction of new technology.

The single incentive fund will be a product of the profitable work of an enterprise, a material expression of its production successes. Its entire staff will be interested in building up this fund as an important source for raising their standard of living. This will be a powerful stimulus for the further development of socialist competition, for the more rapid mastery of advanced experience, and for the pulling up of lagging enterprises to the level of the advanced ones. Lenin attached great importance to applying the principle of rewarding the personnel of enterprises materially for the results of their work, regarding this as an active means of constantly increasing production. Lenin said: it is necessary that "comparison of the business results of the various communes become a matter of general interest and study, and that the outstanding communes be rewarded immediately (by reducing the working day, raising remuneration, placing a larger amount of cultural or esthetic benefits and values at their disposal, etc.)" (Soch., Vol. 27, p. 232).

Let us use our earlier calculations of the norm of profitability for years of the five-year period in order to explain a possible procedure for forming the incentive fund. At present, machine-building factories that fulfill their plans pay their technical and engineering personnel, under the progressive-bonus system, an average of about 10% of basic salaries, and this comprises approximately 2.5% of the total wage fund. Considering the necessity of paying bonuses for technical improvements from a single fund, let us assume that the incentive fund should comprise at least 5% of an enterprise's total wage fund.

Let us put this lower limit in relation to the long-term norm of profitability calculated earlier.

For the first year of the five-year period, the incentive fund should get:

$\frac{0.05 \times 0.30}{0.05} \times 100 = 30\%$ of the profit, where 0.05 in the numerator denotes 5% of the wage fund, 0.30 in the numerator — the share of wages and salaries in production costs, and 0.05 in the denominator — the profitability norm of 5% for the first year of the five-year period for the given type of enterprise.

Accordingly, the percentage payments into the incentive fund from profits in the following years will be:

second year: $\frac{0.05 \times 0.30}{0.09} \times 100 = 16.8\%$;

third year: $\frac{0.05 \times 0.30}{0.13} \times 100 = 11.5\%$;

fourth year: $\frac{0.05 \times 0.30}{0.17} \times 100 = 8.8\%$;

fifth year: $\frac{0.05 \times 0.30}{0.22} \times 100 = 6.7\%$.

It is easy to explain why the payments into the incentive fund from profits decline from year to year: for the amount of payment into the incentive fund to be the same (while the rate of profitability goes up), the rate of payment must be reduced correspondingly. Hence, in order only to maintain a minimum incentive fund at the level of 5% of the wage fund, an enterprise will have to raise its profitability from year to year, i.e., raise labor productivity and reduce production costs. Otherwise it will be unable to pay in full the monthly and yearly bonuses. What the enterprise achieves in any year above the profitability norm, set in advance for the given year, should in large measure go into the incentive fund (for example, 40% of the profit obtained from raising profitability above the norm). Of course, the upper limit of the incentive fund should be restricted to, say, 15 to 20% of the total wage fund, and the maximum size of individual bonuses should also be determined.

The detailed elaboration of procedures for building up incentive funds, the establishment of their maximum size and of methods for their distribution — these are matters of concrete planning. But the principle, in our view, should be that material reward is realized out of net income and in relation to its growth, on the basis of long-term norms set in advance, such as the profitability norm as a generalizing gauge of the effectiveness of outlays of labor under the conditions of the operation of the law of value in the socialist economy.

***

The basic approach should be to elevate planning to a scientific level, relying on the use of the objective laws of the development of socialist production. If we recognize that these laws exist, we are obligated to study and use them. It is inadmissible to recognize them in words and in practice to reduce planning at the center and in the localities to compiling huge statements in which the quantity of output for each line of the list, which often has thousands of items, is with tedious monotony multiplied by quantitative norms, by prices, etc., etc. If only such presentations submitted by enterprises ensured "accuracy" in calculating plan limits! In actual fact, each line may contain some error. Subsequently, the output specifications have to be revised; the norms are often misleading — when they are outdated and when they conform to the practices of the given enterprise and not to what can objectively be achieved by drawing on advanced experience.

Some planners at enterprises are well aware that there is no better means for concealing reserves than an "accurate substantiation" of the plan in bulky, detailed statements. No one can check them, and there is no scientific basis for checking such details. We cannot look for laws of development in the data of an individual enterprise, in its report "base," or in the calculations of its presentations. Laws of development can be found only by studying the indices of aggregates of enterprises, by comparing these indices, grouping them and finding type characteristics, as we have illustrated above. We make poor use of statistics, as was noted at the 20th CPSU Congress. We artificially restrict the comparability of enterprises, supposing that they are comparable only when they manufacture exactly the same goods. This, however, means to descend to technical comparisons, whereas the task is to rise to the level of economic comparisons. Enterprises are comparable within a branch even if they produce a different assortment of goods. They may belong to the same economic type if, for example, the organic structure of their assets is similar and if the average length of their production cycle is approximately the same; and that is precisely what is manifested in such a characteristic factor as the share of wages and salaries in production costs.

From the economic point of view it is important to know not only what goods are produced, but also the objective conditions in which they are produced: the amount of equipment per unit of labor, the structure of the production assets, and, hence, the stable structure of expenditures on production. Where these conditions are equal, society has the right to demand of enterprises of the same type the same degree of effectiveness. This is one of the possible ways of consolidated, simple, and yet scientifically substantiated planning of qualitative indices of production. One cannot, of course, deny the usefulness and necessity of detailed technical calculations in many cases. But as we have already noted, such calculations cannot be taken as a point of departure in planning. What can be taken as a basis is a generalized economic target, a directive expressing law-governed development. And it is the business of the enterprise to substantiate the fulfillment of such a directive by whatever calculations and detailed specifications are necessary for its internal purposes, for its technical, production and financial plans.

By studying the economic indices of the work of enterprises, comparing them with each other, and grouping them by types, we can ascertain exactly what is law-governed with respect to economic effectiveness for each type of production. For example, we can ascertain what should be the profitability of factories at a given rate of growth of labor productivity in the same type of production. The law-governed elements here will be not the average data within given types of enterprises, but the progressive indices achieved by the advanced enterprises.

The task of scientifically substantiated planning consists precisely in separating the objectively necessary factors from accidental phenomena. In particular, the objectively possible effectiveness of the work of enterprises should be separated from accidental phenomena caused by the inefficiency of some managers. In resolving this major task, we can orient ourselves by the advanced specimens of similar enterprises, and we can say: this is the law-governed development, whereas the lag of the other enterprises is a manifestation of the subjective character of management or of other accidental, abnormal factors which the state cannot take into account.

But while demanding of all enterprises the effectiveness achieved by the best of them, we must give them leeway for advancing to the highest level; this leeway is ensured by long-term norms which heighten the personal material interest of enterprise personnel in the technical and organizational advancement of production.

Along with the theoretical elaboration of the questions raised here, and the consideration of different ways of solving them, they should be experimentally checked at a number of enterprises with a sufficiently high degree of specialization. On this road, using the lever of personal material incentive, we will find reliable methods of raising the effectiveness of the work of enterprises.

The proper coordination of planning and material incentive is a guarantee that the decisions of the 20th CPSU Congress on simplifying production management and enhancing the material stimuli of industrial activity will be successfully carried out.

Kommunist, 1959, No. 1

Economic Levers for Fulfilling the Plan for Soviet Industry

E. G. Liberman

The prospects in the building of a communist society, as revealed in the theses of the report by Comrade N. S. Khrushchev at the 21st Party Congress, are tremendously inspiring. Every Soviet citizen wishes to effectuate the Seven-Year Plan more rapidly and, toward this end, to work more efficiently and energetically. The most important factor in the fulfillment of the plan is the selfless and creative work of all working people under the mobilizing and purposeful leadership of the Communist Party. In order to achieve and surpass the rates of growth of production and labor productivity that were fixed by the control figures of the Seven-Year Plan, it is important to direct this activity toward achieving a qualitative improvement of methods of production and a maximum increase in its efficiency.

***

The reorganization of industrial management on the basis of consolidating and developing the principle of democratic centralism and the improvement of planning and the supply of materials and equipment have already brought beneficial results. They have enhanced significantly the creative activity of the personnel of enterprises and have enabled the economic councils to achieve substantial successes in the area of the specialization and cooperation of the enterprises and the fuller utilization of reserves of equipment, materials, and labor. But there are still many tasks to solve. The tremendous potentialities and advantages of the territorial system of management — which has brought leadership closer to production — have by no means been fully realized. To turn these opportunities into reality as quickly as possible, it will be necessary to raise the level of economic work both at the center and in the localities, i.e., in the enterprises, economic councils, state planning committees of the union republics, and the State

Planning Committee of the USSR.

Even after the reorganization of economic management, we still have a complex, multistage system of planning and supply. The supply system is operated through the chief administrations for marketing of the union republics. The existing system of processing orders for the delivery of materials and manufactured items — through the state planning committees to the marketing administrations of the union republics, and then, depending on the nomenclature, through the USSR State Planning Committee to the all-Union marketing administrations — has increased the time required to process the papers involved and has made the task more complicated. For instance, the Kharkov Economic Council did not possess, by the middle of December 1958, the orders and specifications, or the list of suppliers, for many materials and manufactured items required for fulfillment of the program for the first quarter of 1959.

There are frequent discrepancies and miscalculations in the mutual provisioning of production units located in various regions. This is true with respect to many items of the most important lists. For example, the Krasnodarsk Compressor Plant, with the approval of the State Planning Committee of the RSFSR, began the series production of ethylene superpressure compressors. It would seem that an order of such importance for the chemical industry would surely have been backed up with everything necessary. The delivery of pumps and electric motors from other plants was promised. However, when the production of the compressors had been mastered, it turned out that there were no pumps and that orders for them had not even been placed. The compressor plant had to make the pumps itself, and this involved high production costs and a delay; as for the electric motors, the plant did not receive them and, of course, could not make them itself. In October 1958, hundreds of ammonium compressors worth 7 million rubles lay in the yard of the plant, since they could not be sent to consumers because certain completing items were lacking; such are the consequences of mistakes in planning in the machine-building section of the State Planning Committee of the RSFSR.

Is it necessary to cite more examples? It seem to us that in dealing with vast lists of mutual deliveries containing thousands of items, the supply and planning bodies are likely to make mistakes and miscalculations. To avoid them, the central bodies — republic and all-Union — should generally be relieved of this work. The supply system should be built on fundamentally different principles. Large deliveries of materials and manufactured items should be based on direct (i.e., bypassing the central republic and all-Union bodies) long-term contractual relationships between the supplier and customer, effected on the basis of cost accounting and approved by the plan. Any unfounded refusals to conclude contracts or to carry out their terms (which still frequently occur, unfortunately) should be quickly examined by central bodies, whose decisions should carry operative force and not be regarded as mere advice.

As to small deliveries, the economic councils should have depots that can provide the enterprises with nontransit batches of materials and standard items in accordance with the principles of commercial transactions. The employees of such depots should be rewarded with bonuses for having complete assortments on hand and for making deliveries promptly. Obviously, it is not so easy or simple to effectuate all this. But it is essential to strive for it, for otherwise the expanding list of materials and manufactured items and the increasing interdistrict cooperation in production will overwhelm the state planning committees and marketing departments irrespective of the size and efficiency of their staffs.

The state planning committees were conceived as the scientific centers for unified centralized planning and not as operative bodies for the regulation of interdistrict and, at times, even interfactory deliveries of individual items on the most extensive lists. To do away with pernicious "centralization of details" and in this way to ensure centralization with respect to the truly important, cardinal matters, it will be necessary to proceed from the premise that the enterprise is the main link in the industrial system. And the enterprise should actually become the basic organ of planning from below, with the working people contributing extensively to the elaboration of plans. A socialist enterprise is not merely a technical-industrial unit; it is a collective of working people, a "social organism," as Marx put it. The lack of regard for the collectives of the advanced enterprises, when they are called upon to take over the long-overdue assignments of inefficient enterprises, is simply remarkable. And the lagging enterprises are often "aided" by the "switching" of their assignments in order to create the impression that all is well with the branch administration of the given economic council. This practice was widely followed by the ministries and, unfortunately, it has not been abandoned by the economic councils.

Three days before the end of the first quarter of 1958, the Iaroslavl peat workers received from the economic council a "corrected" plan which had been enlarged. As a result, an advanced enterprise was turned overnight into a "lagging" enterprise. The same thing occurred in the following quarter. In April the peat workers fulfilled their plan target by 117.5%. In May they delivered some 2,000 tons in excess of the plan, but one day before the quarter elapsed they received a new plan which again turned the peatery into a "lagging" enterprise. The very same economic council "rounded off" the plan of construction of an oil refinery four times in the first quarter in order to present a favorable figure with respect to fulfillment — 103%. A number of machine-building plants were suddenly converted from lagging into advanced enterprises a day before the month ended. The State Planning Committee of the Ukraine made so many changes in the plans of the Kharkov Economic Council throughout 1958 that the latter was repeatedly forced to change the production programs of the enterprises. This produced a lack of coordination in the material supply service, nonfulfillment of the output assortment plan, and, of course, an abundance of correspondence.

Obviously, such practices undermine the respect of the collectives for the plan as law. More than that, the managements of the enterprises become accustomed to handouts from the planners. At the same time, managers begin to behave with extreme caution; fearing to fall into the "lagging" category through no fault of their own, they often do not disclose their reserves so as to have them available in case they receive unexpected assignments.

The enterprise, which is the fundamental and decisive link in our industrial system, should be given stable, long-term economic prospects. In 1956 it was suggested that economic work should be based on long-term norms. Many agreed with this, but some comrades thought that it would be impossible to know in advance what sort of work the enterprise would be doing over such a prolonged period as five years. They said, among other things, that production techniques were advancing too rapidly for this. Since then the principle of long-term planning has been adopted as the basis for all planning work; long-term plans are drawn up not as reference points for each industrial branch as a whole, but for each enterprise for seven years. This means that it is not only possible but necessary to plan the development of the enterprise for seven years. Unless this is done, it will be impossible to create lasting interplant and interdistrict economic ties.

The problem is to know how to combine plan stability with the necessary flexibility, with the possibility of making changes necessitated by the actual requirements of scientific and technological progress and the appearance of new needs. What can and should be regarded as stable and what should be amended each year as something that requires change?

We maintain that it is possible to foresee five or seven years ahead the type of the enterprise, the general line of specialization and, hence, its further mechanization and automation. Technological progress will necessitate yearly amendments in the assortment of products and, in part, in the volume of output. The production program indices can and should be amended every year. However, the basic qualitative norms can and should remain stable for a considerable period of time. Stability does not mean that they should be identical each year. On the contrary, the norms should be improved progressively, i.e., the dynamics of these indices should be foreseen in advance and registered for each year in the long-term plan. These qualitative norms should not be subject to revision when the yearly amendments to the volume of output and assortment of products are introduced, because the type of production and the main trend of its specialization remain unchanged.

Which indices pertain to these qualitative norms? In our view, the following do:

1. The index of the utilization of means and objects of labor in the form of output per ruble of production assets — both fixed and circulating.

2. The index of labor productivity in the form of output per worker (or per employee).

3. The index of profitability of production in the form of the index of production costs (adopted in our country since 1957) per ruble of commodity produce, measured in fixed wholesale prices.

Would it be possible to foresee the dynamics of these indices a few years ahead irrespective of the yearly amendments in the volume and assortment of produce? We believe that it is possible, as our planning is economic planning and not arithmetical planning. In such a case an enterprise is given an assignment depending on its type and on the basis of an economic analysis of the work of similar enterprises. The assignment would be to ensure improvement in the quality of work each year of the seven-year plan period. For this purpose the enterprise itself would have to design and introduce the production technique that would ensure each year the required level of labor productivity and production costs irrespective of the items manufactured by the enterprise in its particular line.

Even today, labor productivity is planned without account of the output assortment. And it would be unthinkable to do otherwise. But if it is possible to plan the growth of labor productivity irrespective of amendments or changes made in the assortment, it would be equally possible to plan both output per unit of assets and outlays per unit of physical volume of output.

Naturally, it is necessary to encourage the enterprise to meet and surpass these qualitative norms. More than that, the enterprise should not "fear" yearly changes in the volume and assortment of its output. Nor should the enterprise conceal its reserves. These things can be avoided if the material interest of the collective is made dependent not only on the degree of fulfillment of annual quantitative assignments which are subject to amendment, but also, and above all, on the extent of fulfillment of the long-term qualitative norms.

Let us, indeed, assume that an enterprise is rewarded for annual increases in output per unit of production assets. Obviously, in such a case the enterprise will not groundlessly object to higher plan assignments every year, for it is precisely bigger plans that promise continuously higher output per unit of production assets and back it up with corresponding supplies of raw materials, fuel, etc. At the same time, the enterprise will not seek superfluous capital investments or build up above-norm reserves of circulating supplies, but will try to utilize the available assets most effectively.

Higher output per employee will play an equally important part if it is fixed as a long-term norm which will increase every year of the seven-year plan period. The collective will try to disclose the reserves of growth of labor productivity each year in order to maintain and even surpass the set rate for the entire seven-year period.

Finally, the reserves for lowering outlays per unit of output will also be liberated from the arresting influences which now prevail over them, for the achievements of each year (and even each quarter) will no longer serve as the base for planning assignments for the following year or quarter. If attainment of the qualitative indices envisaged in the long-term plan is made the basis for material encouragement, the personnel of the enterprise will have an incentive to achieve a maximum reduction of production costs each year and, hence, to introduce new production techniques.

Why did we take precisely these three indices? Would it be possible to reduce them to only one? Or, perhaps, would it be better to increase their number? Of course, the solutions for all branches of industry cannot be identical.

For instance, in machine building the main index should be outlays per ruble of output (i.e., the index of profitability), for in this branch an above-plan increase of output may be restricted by limited metal resources, and sometimes such an increase is not even needed. An analysis of the work of 25 machine-building plants in Kharkov in the 1955-1957 period revealed that, despite great shifts in assortment, the outlays per ruble of commodity output in wholesale prices fell rather regularly both in the case of types of enterprises and as regards separate plants.

In the extractive and raw material industries, the main role should be played by the index of output per ruble of funds invested in production.

Thus, these indices, in one combination or another, may be applied to any branch of production.

The wage index was not included among the long-term norms because wages should be amended annually in accordance with the actual growth of labor productivity and also with account of the need to plan the circulation of goods and money. But this is a debatable question.

Will not these norms, particularly the indices of output per employee and of output per ruble of production assets, be influenced by changes in the level of cooperation? There is no doubt that they will, but the local body concerned, the economic council, will always be in a position to take them into consideration and correspondingly amend the report indices.

Finally, will not these norms encourage the enterprises to inflate wholesale prices? They certainly will. But all indices will suffer from unsound wholesale prices unless we learn how to overcome such shortcomings. The functions of the all-Union, republic and local bodies in the sphere of price formation are now being delimited, and this measure will facilitate the control of prices. Actually, it is the prices for new and unique products, as well as the temporary prices for newly introduced items before they are put into serial production, that are most vulnerable to inflation. But the number of such prices appearing

every year is not so great that it would be impossible to subject them to strict control with the help of suppliers and
customers.

It would also be useful to create a real material incentive
for the suppliers to calculate the production cost and price of
new items at the lowest possible level. Toward this end, it
would be desirable to allocate to a single incentive fund of the
supplier a sizable part of the sum saved by the customer as a
result of the use of the new product. If the new item is a machine, the economy derived should be calculated on the basis of
capital investment saved. If the new product is an object of
labor, the economy derived will affect the production cost of the
customer's product.

Why do we insist on the need to display a new approach to
the question of material incentives in industry?

It is clear to those engaged in practical work that material
incentives are a potent force, but that this force is not utilized
adequately. Therefore, as soon as a new vital need arises, it is
immediately proposed to create a special incentive system to
satisfy the given isolated need as soon as possible. This practice resulted in the imposition of one system of bonuses upon
another — for fulfillment of the plans for output volume, for the
manufacture of spare parts for tractors and other farm machinery, for the output of new and important products, for the
introduction of new equipment in the machine-building industry,
for the production of consumer goods out of waste materials,
and for many other things — all this in addition to the enterprise fund and bonuses for achievement in socialist emulation.
On the whole, quite a lot of money is spent to provide material
incentives, but we cannot be sure in all cases that production
efficiency is actually being furthered. Each bonus system is controlled centrally by special regulations, and each system has its
own sources of funds. As a result, it often happens that one and
the same person may receive several bonuses for one and the
same achievement, and sometimes for no special achievement
at all.

Despite these numerous forms of bonuses, some enterprises make no effort to further technical progress, put up with
obsolete production techniques, seek to conceal their actual
capacities, and often draft plans which call for excessive administrative personnel and above-norm outlays of materials,
labor, and money. After all, if such norms are included in the
plan, it will be easy for the enterprise to fulfill them, to be

numbered among the advanced enterprises, and to obtain
bonuses. Obviously, such cases are distortions of the socialist
principle of labor remuneration.

It is perfectly clear that the multiplicity of sources of encouragement for various, often overlapping indices should be
abolished. The incentive system should be based on the enterprise as a definite collective of workers. The enterprise should
have a unified incentive fund of its own. It should be able to decide, within definite limits and norms, whom to reward, for
what, and how much; it should be authorized to award extra-
ordinary grants periodically. The collective of workers of an
enterprise, the management, the party organization, and the
trade union committee know best how people work; they know
which people and which efforts deserve to be rewarded. Such a
system of encouragement would make it possible to combine
more fully the personal and social interests of enterprise personnel, the material interest of each employee and a conscious,
communist attitude toward labor. This is particularly important
today in view of the increasingly widespread competition for the
title of "Communist Labor Team." Each worker will know that
his earnings depend solely on his contribution to the labor of the
given collective. The local economic council can exercise control over the distribution of the sums, guided by the size of the
common incentive fund and the maximum size of individual
bonuses relative to base pay.

Therefore, the main object is not to draw up special instructions on bonuses for each new achievement. The primary
need is to establish the principles upon which the unified incentive fund of the enterprise should be formed. The fund should be
large enough to handle all forms of material incentive for the
workers (through the foreman's fund) and office employees. The
enterprise fund should be set up in strict accord with long-term
planning, i.e., it should help simplify planning and, at the same
time, step up plan assignments and increase actual production
efficiency. To secure this goal it will be necessary to divorce
the size of the bonus from the degree of fulfillment of the amended annual plans. Fulfillment of the annual plans for output
volume and assortment is an indispensable condition for the
awarding of bonuses, but the size of the bonus should be made
dependent on actual progress achieved, and not on the accidental
"advantageousness" or "disadvantageousness" of annual plans.
As we have suggested, under the present conditions of long-term
planning, long-term qualitative norms in one combination or

another, depending on the given branch of industry, may serve as the measures of progress achieved. These norms should be worked out on an objective basis, such as analysis and comparison of the efficiency of similar enterprises, irrespective of the economic areas in which they are located.

***

For many years, and not without success, we have been working to reduce and lower the cost of the administrative apparatus. However, there are still some important problems which have to be solved. The fact is that we have not fully succeeded in removing the barriers created by the departmental system of management, barriers that curb the employment and development of local initiative in order to improve the administrative structure. The theses of Comrade Khrushchev's report to the 21st Party Congress attach great importance to this task: "We are not talking about the current mechanical reduction of a part of the existing apparatus, but of far-reaching measures to simplify substantially the structure of the managerial apparatus both at the center and in the localities so as to make it more efficient, competent, and economical."

It seems to us that one of these far-reaching measures should be the increased use of economic levers in influencing production, which would give the enterprises themselves an interest in revising outdated and inefficient forms of management.

Take, for instance, the sphere of labor and wages. The difficulty involved in planning, recording and regulating this vital aspect of production stems from the obsolescence of several categories that were well-founded in the past, but which have lost their meaning today.

For example, we divide the personnel engaged in production into seven categories: workers, engineers and technicians, office employees, junior service personnel, apprentices, watchmen, and firemen.

Is it still necessary to plan and record for all these categories? We think it highly doubtful. At one time there was a good reason for subdividing administrative personnel into engineers and technicians, on the one hand, and office employees on the other, because specialists with a technical education were entitled to certain privileges as regards rations and remuneration. But today this division is outdated, and if the local bodies of the Staffs Administration of the USSR Ministry of Finance exercise their control function in a perfunctory manner it is even harmful: "office employees" are regarded as an undesirable element which should be cut down as much as possible, whereas the category of engineers and technicians is to some extent guaranteed against such an approach. Hence, there is a tendency artificially to inflate the staff of engineers and technicians. It is a fact that to ensure an adequate number of copyists for an enterprise, it is necessary to list some of them as design technicians. One cannot merely hire clerks; it is necessary to elevate them to the rank of "dispatcher" or "technician" and to pay them more. And this by no means leads to lower administrative expenditures.

Such a policy tends to create the dangerous illusion that the relative growth of the number of engineers and technicians is caused by improved organization of production. In reality, the situation is just the reverse: in well-organized enterprises, the share of engineers and technicians in relation to the number of workers is decreasing. Thus, the percentage of engineers and technicians employed at a group of machine-building plants in Kharkov in the 1955-1957 period was as follows: mass production enterprises — 12%; serial production enterprises — 17%; single-unit production enterprises — 25%.

In our view there is no reason to retain the seven categories in taking stock of the working people in industry. Junior service personnel, who comprise less than 1% of the personnel employed in industry, need not be singled out at all. The messengers and cleaners can be listed as workers, while the others can be treated as office employees. The same goes for the watchmen. Actually, from the socio-economic point of view there are only two categories of working people: workers and office employees, and it is only these that should concern the planners and statisticians. The elimination of superfluous categories will greatly simplify planning and accounting work from top to bottom. Hence, it is one of the far-reaching measures for lowering the costs of administering production.

Until recently, workers' qualifications were established in accordance with multi-category tariff scales: in the machine-building industry there were eight categories, whereas in other industries there were ten or more categories. Meanwhile, the machine-building plants never actually apply more than four or five categories. The superfluous categories only make it more difficult to draw up job classification manuals and to establish work norms. Moreover, the multi-category scale entails an excessively large gap between the pay of the top and bottom categories (2.8-fold). The State Committee on Questions of

Labor and Wages has now adopted a six-category scale with a
spread of 1:2. But this is apparently an intermediate step toward the adoption of a four-category scale; world experience
has demonstrated that it is easier, simpler, and more precise
to subdivide labor into four categories which take account of
both training and experience:

    1) untrained workers (including apprentices);

    2) slightly skilled workers (trained but without adequate
experience);

    3) skilled workers (with service and experience);

    4) highly skilled workers (with considerable experience and
an inventive approach).

Reducing the number of categories is another radical measure which should be employed in an effective, and not mechanical, cut in administrative personnel. This can be illustrated
by the following example. At present our plants where the individual piecework system of remuneration is adopted do double
work on a massive scale. First, the rate setters establish the
time norm for each operation, and this norm is then converted
into money. In other words, the rate is established by multiplying minute-norms by the rate of payment per minute for the
given category. The fulfillment of operations by the workers is
also registered twice with respect to the orders issued to them.
The number of hour-norms fulfilled and the amount of money
earned are determined; this is done by double multiplication
of time norms and, separately, of rates by the number of elements manufactured. The results for the month are likewise
summed up in two evaluations — the number of hour-norms fulfilled by each worker and the sum of money he has earned.

Double registration was introduced because there are times
when a worker does a job not in his own skill category, but in a
different (usually higher) category. But if the number of categories is reduced, such a situation will not develop. Each
worker will always be paid in accordance with his own category,
no matter what job he performs. And this means that there will
no longer be any need to do all the calculations in a double
evaluation.

In our conditions, piecework payment remains the main
form of labor remuneration. But is individual piecework remuneration in production still necessary on the scale that some
attempt to apply it, even when it contradicts the conditions of
production? The theory that individual piecework payment best
conforms to the socialist principle of distribution prevails to

this day. But it is not only this form of payment that is in accord with the socialist principle of distribution. Often individual piecework that is unnecessarily adopted for auxiliary operations gives rise, in the case of transport workers, for example, to the desire to carry the same goods back and forth to "produce" more ton-kilometers. Tool shops, in order to safeguard skilled workers against idle time and to guarantee them adequate earnings, begin to put out accessories which are unnecessary.

Even in the main line of production, individual piecework often fails to justify itself. This is particularly true of the conveyor system, where coordination of the work of the entire staff is of paramount importance. Individual incentives entail disturbance of the harmonious rhythm of the line, create disproportions in the form of incomplete parts, cause difficulties in the registration of output, lead to exaggerated output reports and to the appearance of concealed, so-called "uneconomical" waste. The workers are well aware of this. The workers at the Leningrad Carburetor Plant who operate the K-44 Carburetor assembly line demanded that individual orders be replaced by collective orders. Contrary to all "theories" about the harmfulness of collective piecework, the effect was most surprising. Before this measure was introduced, the waste ran up to 20%. Now, thanks to collective responsibility, waste has disappeared altogether. The number of workers on the assembly line was reduced from 15 to 11, and the latter assemble 900 carburetors per shift instead of 800. As a result, all 10 conveyors were shifted to the team method. Whereas before it was necessary to "complete" three or four orders every day, now one order occupies the workers for a whole week. Several of the larger plants in Kharkov, such as the Kharkov Tractor Plant, the Hammer and Sickle Plant, and the Bicycle Plant, have adopted the method of collective payment for the finished product of sections.

Experience has shown that collective payment actually increases the material interest of the workers in the results of their labor, meaning by results not only individual output, but also fulfillment of all the plan indices by the team or section.

The principle of team organization is just as effective in sections engaged in small-scale serial production. The Stalingrad Drilling Equipment Plant has introduced the team method. This step has strengthened collectivism in labor and has made it possible to hand over the working place to the relief without discontinuing work on units. In the past, the change of shifts took from 20 to 30 minutes, which were wasted. Each worker used his own tools, the quality of work was different, and the equipment poorly maintained. Today there are no lagging workers in the teams.

It is apparent that the principle of individual piecework should be revised in many cases, for it does not always correspond to the collective character of labor in many sectors of production.

The time is ripe not only for adopting collective forms of remuneration, but also, in some cases, for introducing time-and-bonus payment on a broader scale. We cannot disregard the fact that piecework remuneration sometimes hinders the introduction of new technology, for the latter involves the downward revision of piece rates. People are often surprised at the fact that wages at some plants have recently been growing faster than labor productivity. The reason is precisely the one indicated above: although new technology is being introduced, neither the shop superintendents, nor the foremen, nor the workers themselves have sufficient inducement for corresponding reductions in the rates of payment. The time-and-bonus system of payment is free from these shortcomings. The more our industry is mechanized and automated, the more reason there is to pay the worker a fixed rate and to grant the teams a collective bonus for carrying out the daily program, maintaining equipment in good order, and working without waste. Time-and-bonus remuneration, like piecework, corresponds to the principle of remuneration of labor in accordance with its quantity and quality, provided it is introduced thoughtfully, in sectors requiring great accuracy or a particularly high level of work, or for auxiliary jobs, for the servicing of systems of automatic machines and conveyor lines.

Many more examples could be cited from the sphere of the organization of labor and wages. But even the above-mentioned cases clearly indicate that the management of labor in production is complicated by unnecessary, outdated and, at times, stereotyped methods of organizing labor. The enterprises are not seriously encouraged to reduce the costs of accounting and clerical work or to simplify and reduce the administrative apparatus. A considerable influence is exerted by perfunctory financial control, which fails to get to the essence of the matter and saves a kopeck for the state at the price of an overexpenditure of thousands of rubles.

As we know, some small enterprises have had good results from eliminating the shop system of management. An enterprise employing up to 500 or even 1,000 workers is able to cut its administrative staff by at least 20 to 30% if it does not have the shop system of management. This does not merely involve the renaming of shops into sections and the mechanical reduction of personnel. It involves a fundamental reorganization, with greater rights and responsibilities for the section foremen.

But the non-shop system is being introduced very slowly. And the main reason for this is that the agencies exercising financial control often fail to consider the essence of the matter. Many of the shop superintendents should, in keeping with the new structure, become section superintendents. And although their responsibilities will actually increase, their rate of pay will immediately decline. Assistant directors of plants often become chiefs of supply departments, and chief engineers become heads of technical departments. Here, too, the essence and volume of their work do not change, but their pay is reduced. Obviously, if the non-shop structure is adopted, considerable savings should be derived from cutting down the number of redundant personnel and not from cutting the rates of pay of those who remain at their jobs.

A curious case may be cited from the field of the consolidation of enterprises. We know that the consolidation of enterprises is one of the radical methods of simplifying administration. After the establishment of the Kharkov Economic Council, the small Kharkov Agricultural Machinery Plant specialized successfully in the manufacture of certain assemblies and elements for the general-purpose diesel engine SMD, produced by the big neighboring Hammer and Sickle Plant. Of course, these plants had to be merged, with the former transformed into a shop of the latter. As a result, it would be possible to cut down the administrative staff by several dozen people. There would no longer be any need for two tool and two repair shops. The correspondence, reciprocal claims in the economic council, and settling of accounts through the State Bank, which create an excessive circulation of documents, would no longer continue. In a word, the case was clear. But there was one obstacle that the personnel of the economic council were unable to overcome because of the existing system of recording gross output by the plant method. As a result of the merger, the economic council would "lose" approximately 100 million rubles worth of output a year. While the Kharkov Agricultural Machinery Plant was a separate enterprise, the assemblies and elements it manufactured were registered twice — once as the output of the said plant and a second time as a component part of the value of the diesel engines put out by the Hammer and Sickle Plant. This output would be included only in the value of the diesel engines if the plants were merged. It follows that the existing system of taking stock of output by the plant method induces the economic council to split enterprises rather than to merge them. This is a typical case of fetishism with respect to rules and schemes that we ourselves have created. A clearly irrational procedure is preserved in the name of an accounting fiction which has virtually acquired the force of law.

All these individual phenomena are indicative of certain general features which stem from the shortcomings of economic work in industry. The majestic goals of the Seven-Year Plan require that we improve this work, that we make much better use of the economic levers for raising the efficiency of production. This is one of the most important conditions for the successful execution of the economic plans during the period of the comprehensive building of a communist society.

The Discussion in 1962

Voprosy ekonomiki, 1962, No. 8

Planning Production and Standards of Long-Term Operation

E. G. Liberman

The new Party Program adopted by the 22nd Congress of the CPSU presents a number of fundamental propositions from which one should proceed when assessing established planning practice. Of these, the following are most essential for the subject under consideration: "The economic independence and the rights of local organs and enterprises will continue to expand within the framework of the single national economic plan. Plans and recommendations made at lower levels, beginning with enterprises, must play an increasing role in planning." "The entire system of planning and assessing the work of central and local organizations, enterprises and collective farms must stimulate their interest in higher plan targets and in the maximum dissemination of advanced production experience."

The present procedure of planning the work of enterprises stifles their initiative, does not permit the maximum utilization of production potentialities and the advantages of the new system of management, and does not make enterprises interested in further raising the efficiency of production. This is suggested by countless facts, by statements of industrial executives and scientists.

How, then, are we to fulfill one of the major requirements of the Party Program, namely, that of making enterprises more interested in increasing production to the maximum and in improving its quality?

In my opinion one must not confine oneself to correcting or supplementing individual plan indices. Only the key indices, the decisive indices, should be handed down to enterprises, whose directors should be given greater rights and opportunities for economic maneuvering within their scope. The quality of an enterprise's work should be assessed on the basis of its ultimate efficiency and not on the basis of a multitude of indices regulating in detail the economic operation of the enterprise.

I support the so-called principle of standards of long-term operation — one of the possible variants of the comprehensive solution of the problems of extending the rights and initiative of enterprises. What are the principal demands that society makes of socialist enterprises? First, maximum output. Hence, the first thing that must be handed down to an enterprise is its assignment for the volume of marketable output of a specified nomenclature and appropriate quality. This should be the principal indicator of the work of an enterprise, the condition for the growth and proportionality of production and the most rapid creation of the material and technical basis of communism. Second, maximum efficiency of production and realization of output. But is it necessary to hand down to enterprises such a large number of indices of efficiency as is now done? Important as many planned and accounting indicators may be, there is no need to impose them upon enterprises from above. Centralized planned guidance should be concentrated mainly on working out and insuring the fulfillment of the key indices of plans, always taking account of proposals coming from below. It is precisely in keeping with this requirement of the Party Program that enterprises should be given the right (and obligation) independently and completely to elaborate technical-production-financial plans on the basis of the output target (with the nomenclature taken into account) planned for them "from above." This means that the indices of growth in labor productivity (and thus in size of work force) in the wage fund and the average wage, in the reduction of production costs and increase in profitability of enterprises should be worked out directly at the enterprises. The enterprises themselves should draft the plans for installing new machinery, for capital investments and for all financial indices.

A legitimate question arises: what guarantee is there that, in the course of the above, enterprises will be primarily concerned with state interests? In our opinion we can and must provide such a guarantee. This is the key to improving and simplifying planning work. It may be provided under the condition that what is profitable to society as a whole will also be profitable to each production collective and, on the other hand, what is wasteful from the standpoint of public interests will be extremely unprofitable to each enterprise. In principle there should be no place under socialism for contradictions between the interests of society and those of an individual enterprise. But various adverse features of our economic management, stemming from still persisting elements of excessive centralization and economic arbitrariness of the period of the personality cult, frequently result in the interests of the national economy and of the personnel of individual enterprises coming into conflict, and in some cases this leads to the concealment of potentialities, to a striving to understate plans, and to indifference toward initiative in rationalization and toward the dissemination of advanced experience.

All this results from the fact that we still are not planning and stimulating production on the basis of true principles of democratic centralism. The reorganization of the managerial machinery, aimed at decentralizing the latter, is meeting with opposition on the part of the old planning system that has been largely retained, planning by the departmental-centralized method of "requests" and "allocations." What is more, this old system has become more complicated in some ways. Inasmuch as the enterprises and economic councils have not been given the necessary independence, a multitude of management levels have appeared, frequently issuing uncoordinated directives. This leads to superfluous work involving the altering and correcting of plans, and takes up much effort and funds.

The requirements set forth by the Party Program oblige us to work out a simple and efficient system of direction devoid of attempts to "urge on" enterprises and disclose their potentialities by paper methods of allotting resources. In our opinion a guarantee of conscientiousness in "planning from below" can be created on the basis of application of the "share-in-the-income" principle: the more values an enterprise has created for society, the larger the sum that must be channeled into its financial incentive fund, regardless of whether these values have been produced within the scope of the plan or over and above it.

The "sharing" principle is realized in the form of a planned long-term standard of profitability of production. Indeed, besides the volume of output in physical terms, only one additional assignment must be handed down to enterprises, namely, that of "operating at maximum efficiency." First of all, efficient operation increases the profitability of production, especially if it is measured as the percent profit on production assets (fixed and circulating). (1) Hence, the index of profitability of an enterprise must be handed down as the second basic assignment of the enterprise. This index should make enterprises unconditionally interested in improving all production indices both in the process of elaborating the plan and in the course of its fulfillment. This may be achieved by drawing up for uniform groups of enterprises within each branch of industry a standard scale of deductions from profits put at the disposal of an enterprise depending upon its level of profitability. There is no need to hand down any other indices to enterprises. To illustrate this idea, we present an approximate scale which we have worked out for branches of the machine-building industry (see page 78).

As the table shows, the higher the profitability (in % of assets), the greater is the amount of financial reward. Thus, as profitability rises from 5.1 to 60%, the maximum size of the financial reward fund of an enterprise increases more than 2.5 times. But at the same time the sum of profits received by the state budget grows much more rapidly, i.e., 18 times, $\left(\dfrac{60 - 5.3}{5.1 - 2.1} \approx 18\right)$. This provides a guarantee of exceedingly high deductions placed at the disposal of an enterprise even when profitability is very high. The scale, of course, is approximate, but it was established on the basis of an analysis of data on the operation of 23 plants over a five-year period and was checked by statistical modeling. The scale of deductions is a tabular expression of the logarithmic function of profitability.

Worked out in a differentiated manner for branches of industry and groups of enterprises, such scales should serve as standards of long-term operation, inasmuch as they will not vary throughout at least the period of operation of a given set of wholesale prices. An enterprise will thus receive a single quantitative assignment for the year — the volume of output of a specified nomenclature, as well as the standard scale of profitability approved for the group of homogeneous enterprises to which it belongs.

The main objection to the practical application of the standards of long-term operation according to profitability of production in the form in which they were proposed in 1956 consisted heretofore of the view that, in conditions of a rapidly developing technology and economy, one could not foresee five or more years ahead what the profitability of this or that group of enterprises would be. But if the standard is given in the form of a scale of deductions from profits, this objection no longer arises, since there is no need to foresee a number of years ahead what the profitability of each plant will be. In our view this scale reflects the possible range of profitability of enterprises belonging to a given homogeneous group. It will depend upon the plant itself how quickly it advances along the scale of financial rewards.

Scale of Financial Rewards for Enterprises Depending Upon Level of Profitability*

Financial Reward	Profitability (in % of assets)						
	Up to 5	5.1-10	10.1-20	20.1-30	30.1-45	45.1-60	60 and higher
In % of funds	—	2.1	3.0	3.9	4.4	4.9	5.3
Additional (in % of profits exceeding the lower limit of the interval)	42.0	18.0	9.0	5.0	3.3	2.7	2.0

*The total sum set aside for financial rewards may be increased or decreased by the following amounts depending upon the output of new (non-comparable) products; decreased by 3.5% when the share of new output is up to 10%; increased by 3% when the share of new output amounts to from 25.1 to 35%; increased by 7% when it amounts to from 35.1 to 50%; increased by 15% when it amounts to from 50.1 to 70%; and increased by 25% when it amounts to 70.1% and higher.

On the basis of its quantitative assignment and guided by the standard scale of profitability, an enterprise must independently draw up its technical-production-financial plan for all the rest of the indicators, which by no means lose their force and significance. An enterprise will try to provide for high profitability in the plan itself and strive to achieve it.

Under the suggested system of planning, an enterprise will receive deductions from profits based on the percentage of profitability included in the plan which the enterprise itself draws up. This means that it will be pointless for an enterprise to artificially reduce its plan. But at the same time the incentive to overfulfill it is retained, since with the growth of profits the size of the financial reward increases, even when the planned rate of deductions is applied. Naturally, if the actual profitability is below that planned, the enterprise will receive financial rewards in accordance with the scale for actual profitability. We should observe the rule that deductions from profits

placed at the disposal of an enterprise must be the single and
sole source of payment of all types of financial rewards: for
fulfillment of plans and for reduction of production costs, for
mastering new machinery; and, what is most important, of
payment of lump-sum premiums by decision of the management and trade union organization of a plant, and also of funds
for above-plan housing construction and other needs of public
consumption. An enterprise will be vitally interested in attaining high profitability both in working out its plan and in
implementing it.

Finally, the factors upon which profitability depends, at
the given planned prices, must be taken into account. These
are, first, full capacity operation and growth in the volume of
production. We know that in the machine-building industry
each 10% growth in production due to intensification results
in no less than a 2.5-3% reduction in production costs and,
thus, in growth in profitability. Hence under the new system
of planning an enterprise has no reason to fear increased output assignments. Second, profitability depends upon the direct
reduction of production costs. And the proposed system of
planning provides the incentive for an enterprise to work for
minimum production costs both in terms of its plan and in
actual fact, for disclosing potentialities for economy and
growth in labor productivity, since the assessment of its work
will depend not upon the plan but upon the actual profitability
expressed as a percentage of assets. In this connection it
loses all desire to draw in additional labor power, increase
the rate of consumption of materials, and obtain unnecessary
equipment and excess capital investments. At present such
"excesses" do not prejudice an assessment of the work of an
enterprise, do not affect the fulfillment of its plan with respect
to many indices. On the contrary, the greater an enterprise's
reserves of equipment, labor power and various supplies, the
easier it is for it to deal with all manner of "surprises" in
planning. Under the proposed system, any idle machine tool
and unused production floor space, each superfluous worker
and redundant stock of materials will decrease the actual size
of profits relative to production assets, will hit the "pocket-
book" of plant personnel as a whole and each employee per-
sonally. It seems to me that where there is a common source
of financial reward the proper combination of individual and
collective material interest in the successful operation of an
enterprise is created, the need for which Comrade Khrushchev

stressed in his report on the Program of the CPSU at the 22nd Party Congress. This common source is provided by the fact that both individual bonuses and all forms of collective bonuses for building up the public consumption funds (kindergartens, nurseries, vacation homes, etc.) will be paid out of one and the same fund.

In drafting a plan an enterprise must, of course, adhere to established nationwide standards, such as official salaries, rates of basic wages for manpower, rates of depreciation, and maximum sizes of personal bonuses. It is possible that the latter should be increased somewhat, but the main thing here is that all the types of bonuses come out of profits and not out of production costs. On the basis of the results for each quarter, the profit and profitability relative to production assets are calculated; from this the size of the deductions for financial rewards for the collective are determined according to the scale. In the following quarter the enterprise will pay out all types of bonuses from this fund. The personnel of an enterprise always has the right to receive a bonus if the output plan, with due regard for the assigned nomenclature, has been fulfilled. But the size of the financial reward is limited by the available fund which the collective has "earned" by its work. The financial reward fund is thus a carry-over fund. To enhance the effect of all cost accounting monetary sanctions (penalties, fines and forfeits) on the economy of an enterprise, they should not be directly charged to production costs but should be paid out of the financial reward fund and, what is more, by decision of the collectives a certain part of these sanctions should be charged to the guilty parties. Under such a system the paradoxical situation in which a poorly working enterprise reduces production costs in comparison with the planned figure by not paying out bonuses stipulated in the plan will be eliminated.

***

Centralized planning of qualitative indicators should be conducted according to the following scheme. On the basis of long-term continuous plans and appropriate balances, the USSR Gosplan [State Planning Committee] and the Gosplans of the republics calculate and determine for each economic administrative area the summary control figures for all necessary qualitative indicators. But the economic councils, guided by the ill-famed "accumulations of reported data," should not mechanically allocate indices among industrial administrations and enterprises in the usual manner. The economic council can and should become the center of economic work, securing to enterprises the conditions for highly profitable activity through specialization, cooperation and the introduction of new machinery. The threads of generalized state targets (control figures) from above and the plans of enterprises from below should come together in the economic council. Since enterprises will be interested in achieving maximum results with minimum outlays, the sum total of the plans of enterprises will fully insure the fulfillment of the summary control targets for the area as a whole, including such key indicators as labor productivity, wages and production costs. Many economists believe that this proposal is naive, to say the least, and will inevitably lead to the practice of Gosplan and the economic council "putting pressure" on an enterprise and the enterprise "pushing back." But any good plant director will confirm that if he is not annually "trimmed down" to the "attained level," if he can obtain sufficient financial reward funds depending upon the profitability of the enterprise's work, if he has a long-term prospect of increases in this profitability without arbitrary changes in the standards of deductions from profits placed at the enterprise's disposal, the plans of enterprises will be sufficiently intensive, since their personnel will themselves be interested in this. As a result, enterprises will work much better and planning will become much simpler and more effective.

Let us even suppose that the plans of enterprises, contrary to their own material interests, will nevertheless be understated for some qualitative indicators. In such cases the economic council should retain the right to demand that the plans be modified. Mathematical methods can be used here to the best advantage, correlation methods in particular. These permit a sufficiently exact estimation of the extent to which indices in the plans of individual enterprises conform to their objective characteristics. However, we should note the big difference between this and the existing procedure, namely, enterprises will not give a hostile reception to any improvement in planned indices suggested by the economic council, since this will not alter the standards of financial reward and will not worsen the position of an enterprise in the assessment of its future work. Let us recall again that if the output plan in its basic nomenclature is fulfilled, an

enterprise receives the right to a financial reward depending
upon the profitability of its work. All the other indicators —
labor productivity, labor force, wage fund, production costs —
are planned and taken into account from bottom to top as
highly important national economic indicators which are obligatory only for the economic councils. But for enterprises
these indicators should not serve as a means of assessment
or a condition for financial reward. Plant directors should
themselves formulate the optimum combination of indices in
plans, lowering some and raising others, so as ultimately to
derive the maximum effect, namely, growth in output, improvement in its quality, and an increase in profitability of
production. This is exactly what freedom of economic maneuvering and initiative from below consist of.

The most flexible means of insuring improvement in
quality of output and its durability is the system of financial
rewards for profitability. If a product offers its user the
possibility of additional applications, a corresponding addition
to its price should be established — this is the conclusion
reached by almost all directors of enterprises. In this way
profitability of production will be increased, which in turn
will fully recoup the increase in outlays of labor and materials connected with improving the quality of the product.
Limiting individual indices of labor productivity and production costs hampers the qualitative improvement of production. For this reason I believe that the suggestions to provide
financial rewards for betterment of the previous year's results should be rejected (in particular, the proposals made
by the executives of the Kirov Plant in Odessa to regard a
reduction in production costs compared with the previous
year as profits). This contradicts the principle of continuity
in planning. On what grounds should each subsequent year
cancel all past achievements and each plant start making
progress "from scratch"? Production is of a continuous
nature and the personnel of each enterprise must be confident
that, having worked earnestly this year on, say, automation,
they will be able to profit by its results for a number of
years to come and not only up to the beginning of the new plan
year. But most important is the fact that, under a system of
financial rewards based on "the attained level," enterprises
will have no incentive to introduce improvements in the design of goods and in the technology of their manufacture.
Aircraft designer Oleg Antonov has written about this. If the

work of an aircraft engine plant is assessed on the basis of
its reduction of production costs compared with the previous
year's results, what sense is there in increasing the working
life of engines? After all, as a rule this increases their production cost. It is a different matter if a somewhat higher
price, let us assume by 10%, is fixed for an engine, taking into
account the fact that one new engine is equivalent to, say, one
and a half old ones. Then the higher outlays on improving output will be fully justified and it will always be profitable for
the plant to improve the quality of output. (2) At the same
time this will be economically advantageous from the standpoint of the user as well, and hence of the national economy
as a whole.

There are a multitude of difficulties and doubts about the
system of financial rewards for profitability. One of the apprehensions expressed, for instance, is that enterprises will
artificially boost prices of new products. But at present the
price of the supplier's goods makes no difference to the plant
doing the buying: the latter demands delivery of the goods at
any price, if only it has been "legally" approved, since subsequently the outlays stemming from the purchase of equipment
at a higher price can be taken into account in the estimate of
capital outlays or in the planning calculations of production
costs, and the plan "justifies" everything. Under the proposed
system of planning, a higher price for any elements of production, be they fixed or current assets, will lower actual
profitability, since the new criterion of an enterprise's activity will have an objective character and will not depend
upon what it is possible to "drag" into the annual plan. Buyers will carefully check the price calculations of the goods.
Control over prices is thus put on the only sound basis —
a cost accounting basis — since a formal check of the mani-
fold calculations in economic councils and in price bureaus
of Gosplans results only in a delay of the work and the enlargement of staffs. In my opinion the central bodies should
approve prices only after they have been mutually checked
by enterprises on the basis of direct ties established between
suppliers and buyers.

Some economists fear that enterprises will find themselves in different circumstances, since profitability supposedly does not depend upon them. But one must look into the
causes of sharp differences in profitability to ascertain
whether these stem from objective conditions which do not

depend upon the enterprises, or are connected with the quality of their work, which depends to a considerable extent upon management and the entire personnel. Objective differences in profitability of enterprises arise because differentiated standards of profitability are in force in different branches of industry. That is why we suggest different (and not uniform, as many authors propose) standards of profitability and deductions for the different branches of production. In some branches differences in profitability are explained by the existence of various rents — land, transport, mining, and others — but these can also be taken into account by grouping enterprises according to similar conditions of work. Differences in profitability within one and the same branch of industry may arise because enterprises are unequally equipped technically, but this too can be taken into account in grouping them. However, the backwardness that an enterprise itself can eliminate should not be justified by reference to standards.

One of the variants of differences in the technical equipment of enterprises is an unequal level of specialization, and this means the scale of serial production, which may prove to be the cause of the different profitability of the same goods at plants producing identical products. But this is no reason to lower the standard of profitability at individual enterprises, which would justify the low level of specialization. On the contrary, it is necessary, with the help of economic tools, to exert influence upon the economic councils and enterprises with a view to correctly distributing output and establishing rational intra-regional and inter-regional cooperation.

Frequently the cause of unequal profitability lies in the fact that one plant is perpetually developing the manufacture of some new product and hence shows a low profit, while others are year in and year out producing a long-since developed product and reaping high profits. In my opinion coefficients of addition to the deductions-from-profits scale can be calculated and established for enterprises developing new products, coefficients which take into account the share of new output in the plant's production program.

We mentioned above that an addition should be made to the prices of new goods which would reflect their higher quality, providing greater usefulness for buyers. As for long-since developed goods, the prices of these should be reduced when their profitability rises above a reasonable limit. The enterprises themselves can also have a stake in this if in the first year after the reduction in prices the state's profit from the reduction is regarded as profit from which corresponding deductions will flow into the fund of the enterprise (this proposal comes from Comrade Stepanov of the Ul'ianovsk Economic Council).

Another cause of fluctuations in profitability is the current prices of industrial goods. But if prices will be determined on the basis of average production costs for a branch of industry plus the same branch rate of profit, profitability will be a quite adequate criterion of an enterprise's efficiency of operation.

Our present planning methods and practice proceed from the premise that the most reliable assessment of the work of enterprises upon which their financial rewards depend is fulfillment of the plan. This is justified by the fact that the plan supposedly creates equal conditions for all enterprises, that it takes into account different natural conditions, unequal levels of mechanization, and other "individual" peculiarities. I consider this to be incorrect. After all, along with actual differences in objective conditions, planning also attempts to equalize conditions such as technical backwardness, which the management easily reconciles itself to if an allowance for this backwardness is legitimized by the plan. But the greatest danger consists in the fact that at present the differences in quality of work are leveled off by planning based on "accumulations of reported data." Behind these lie differences in the initiative of personnel, the level of development of socialist emulation — that is, everything that depends entirely upon human effort. Thus, no "equal conditions" whatsoever are created, since the plans of enterprises, with the aim of creating "identical conditions," are drawn up primarily by taking into account the previous level reached by an enterprise. Thus, more favorable conditions are created for poorly working enterprises or enterprises which were able to secure more advantageous plans for themselves, and disadvantageous conditions for active collectives who work well and fully disclose their reserves.

This concept — "equal conditions through plans" — leads to leveling in the quality of management of enterprises. Why strive for good work if it is simpler to try to get a "good" plan? In our opinion it is high time to reject such a concept. The origin of the concept is historically explicable: at one time any growth in production was effective. Hence the appearance of the theory which held sway until recently and was denounced at the 20th Congress of the CPSU to the effect that everything is advantageous under conditions of socialism. Since there is no competition, neither is there, supposedly, any obsolescence; hence both out-of-date and modern enterprises can work equally well. But obsolescence in the sense of aging from the economic standpoint is characteristic not only of equipment. This category should also be extended to embrace outmoded methods of production and systems of management which were previously acceptable but now hamper development of the national economy. However, if this aging is concealed by a plan adjusted to "individual conditions," it can never be economically discredited. After all, it is one step from this concept to the assertion that the economic efficiency of operation of enterprises is not an objective category and is created, as it were, by ourselves: if we submit an understated plan, then an enterprise with obsolete equipment or an enterprise with a deep-rooted bad style of work will appear to operate quite efficiently; if we raise the plan of a leading, technically well-equipped enterprise the latter will, on the contrary, appear to be "bad."

Really "equal conditions" can be created when we have a single standard for enterprises that are in approximately the same natural conditions. There is less danger in ignoring certain differences in objective conditions than in leveling the quality of management. The Plenary Meeting of the Central Committee of the CPSU and the zonal conferences of agricultural personnel held after the 22nd Party Congress have once again shown how different the results can be for farms in identical objective conditions. These differences are determined entirely by the quality of management. Hence it follows that attempts to take into account individual conditions through the plan are for the most part harmful. By such leveling, backward methods of production are preserved and, what is especially dangerous, the actual worth of our executives is concealed. Enterprises, upon receiving a production program and a long-term standard of profitability from the center, must work for the best results. There is no anarchy and competition in our economy but that does not at all mean that there is also no emulation designed to achieve the best methods of management. On the contrary, such emulation should receive proper scope.

* * *

The suggested scheme is, of course, not the only possible one. Neither is it a panacea for all shortcomings. This scheme, or possibly more successful similar ones, can only serve as an important aid in improving the organizational and educational work of Party and government bodies.

When personnel of our central economic apparatus are faced with the problem of eliminating various shortcomings in planning practice, they frequently solve them by introducing certain new planning indices. But the indices by themselves will remain hanging in mid-air if material incentives for their fulfillment are not provided. It is necessary to introduce special bonus regulations providing financial rewards for fulfillment of indices. These regulations inevitably become overgrown with instructions and special levels of control are needed to protect state interests from possible abuses. But this does not achieve its purpose since personnel capable of "squaring things" and "pushing things through" more often than not get around the barrier of instructions. This occurs because distrust of enterprises lies at the basis of lower-level planning. Centralized planning can be strengthened and simplified if the need for material incentives for enterprises in the planning process itself is taken into account. That is why it seems to us that, no matter what the proposals for radical improvement of planning in industry, they must rest on long-term standards of profitability of enterprises based on the utilization of material incentives to stimulate interest in improving planning and increasing the efficiency of socialist production.

Footnotes

1) Relating profits to production assets is only a way of measuring production efficiency. This does not mean that the scheme of "price of production" should be taken as a must in the process of price formation.

2) See Izvestia, Nov. 18, 1961.

Pravda, September 9, 1962

Plan, Profits, Bonuses

E. G. Liberman

It is necessary to find a sufficiently simple and, at the same time, substantiated solution for one of the major tasks set by the CPSU Program, which is to develop a system of planning and assessing the work of enterprises so that they will be vitally interested in the highest possible plan targets, in introducing new machinery and improving the quality of output, in a word, in the highest efficiency of production.

This can be achieved, in our opinion, if the enterprises are presented with plans only with respect to the volume of output and the assortment and dates of deliveries. Moreover, this should be done in such a way that the direct contacts between suppliers and consumers are taken into account as much as possible.

All the other indices should be communicated only to the economic councils; there is no need to distribute them among the enterprises.

On the basis of the targets for volume and assortment of output, the enterprises themselves should work out a complete plan that covers, among other things, labor productivity, quantity of work force, wages, costs of production, accumulations, capital investments, and new machinery.

How can the enterprises be entrusted with the job of working out plans when at present all their draft targets are usually much lower than their actual capacities?

This can be done if the enterprises have a maximum interest, both moral and material, in making full use of their reserves not only in the process of the fulfillment of plans, but also during the drafting stage. For this purpose it is necessary to work out and approve long-term plan norms of profitability for every branch of production. It would be most expedient to approve these norms centrally, in the form of scales determining the size of bonuses to enterprise staffs in dependence on the attained level of profitability (in the form of profit in percent of production assets).

On the following page is an approximate scale of bonuses worked out for machine-building enterprises on the basis of analysis of the work of 24 plants in five years.

Here is an example of how the scale is used. Let us assume that an enterprise's balance profit for a year is 7.5 million rubles, with the average annual sum of its fixed and circulating assets amounting to 50 million rubles. This means that the enterprise's profitability is 15%. In this case the enterprise has the right to receive as encouragement bonuses, first, 3 kopecks per ruble of the assets, i.e., 1.5 million rubles; and second, 9% of the sum of the profit above the lower limit of the interval, i.e., above the 5.05 million rubles which make up 10.1% of the value of the assets. In the given case we take 9% of 2.45 million rubles, or 221,000 rubles. Thus, the enterprise will get a total of 1,721,000 rubles.

The rates of the scale can, of course, be either increased or reduced. The important thing at this point is the principle and not the magnitudes.

The principle is, first of all, that the higher the profitability, the greater the reward. For instance, when profitability rises from 5.1% to 61%, i.e., twelve times, the reward to the enterprise increases from 2.1 kopecks to 5.3 kopecks, that is, 2.5 times. This guarantees that the enterprise will have a strong material incentive to increase profitability. At the same time, the sum of the revenue going to the state budget rises much more rapidly: correspondingly, from 3 kopecks to 54.7 kopecks per ruble of the assets, i.e., 18 times. This guarantees an even more rapid growth of social wealth and, at the same time, insures against inordinately high deductions in favor of the enterprise. There is absolutely no danger for the budget revenues: on the contrary, there are grounds for expecting a substantial increase in state income under the influence of the enterprises' strong material interest in the general growth of profit.

Second, the principle envisages that the enterprises will get bonuses on the basis of their share of participation in the income created: the greater the profitability in the plan, which is compiled by the enterprise itself, the greater will be the bonus. If the profitability plan is not fulfilled, the enterprise will be rewarded according to the same scale, depending on its actual profitability. If the profitability plan is overfulfilled, the

Approximate Scale of Bonuses for Enterprises

Profitability (balance profit in % of the fixed and circulating assets)	0.01 to 5.0	5.1 to 10.0	10.1 to 20.0	20.1 to 30.0	30.1 to 45.0	45.1 to 60.0	Over 60.0
Bonuses a) in kopecks per ruble of assets	0.0	2.1	3.0	3.9	4.4	4.9	5.3
b) additional bonuses in % of the profit above the lower limit of the interval	42.0	18.0	9.0	5.0	3.3	2.7	2.0 *

* In all, not more than 5.5 kopecks per ruble of assets.

enterprise will receive bonuses also according to the same scale, but on the basis of the average rate of the planned and the actual profitability. Let us suppose that the enterprise plans its profitability at 10%, and that it actually reaches 15%. Then its bonus is based on the profitability rate of 12.5%, which is the average of the planned and actual profitability. This means that it will be extremely unprofitable for the enterprise to work out plans below its capacity. At the same time, the incentive for the overfulfillment of the plan will be retained. Naturally, if the enterprise does not carry out its assignments with respect to the volume, assortment, and delivery dates of output, it will be deprived of the right to any bonus.

On this basis it is possible to simplify considerably and to improve local planning. The fact is that the enterprise will get only one assignment for the output of goods with respect to assortment and, besides, it will proceed in its plans from a set scale of reward for profitability.

In order to achieve high profitability, the enterprise, as it works out its plans under our conditions and with the plan prices, will have to strive for the fullest possible employment of capacities and equipment (for profit will be calculated in percent of the assets). This means that the enterprise, in pursuance of its own interests, will increase the shift index and load of its equipment and will stop asking for superfluous capital investments and machine tools, that it will not create unnecessary stocks. Whereas at present all these superfluous stocks serve almost as a gratuitous reserve for the enterprise, under the new system they will be painfully felt financially, for they will reduce the size of the bonuses. Consequently, the enterprise will no longer "fight" to get unjustifiably low plans, for such plans will not give the enterprise sufficiently high profitability.

Moreover, the enterprise will strive for a maximum reduction of the cost of production; it will be thrifty and will not artificially overstate the norms for the consumption of materials, fuel, tools, and power. The fact is that the reduction of expenditures will result directly in the growth of profitability, regardless of the norms that can be "railroaded" into the plans and orders. Consequently, profitability in percent of the enterprise's fixed and circulating assets is an objective criterion: it does not depend on what kind of a plan the enterprise tries to secure for itself.

Finally, and this is the main point, the enterprises will strive to raise their labor productivity to the maximum. They will stop asking for, and hiring, superfluous manpower, for this will reduce profitability and, consequently, the incentive funds.

The incentive fund, formed in accordance with the level of profitability achieved, should be a uniform and exclusive source of all kinds of bonuses. It is important to consolidate collective and individual material incentives. Let the enterprise have more freedom in using "its own" share of profit!

The question naturally arises: will this maintain and strengthen the principle of centralized guidance in our planning?

There is every reason to assert that the system proposed will relieve centralized planning of petty tutelage over enterprises, of the expensive attempts to influence production not by economic measures, but by administrative techniques. The enterprise itself best knows its reserves and can find them. But in doing so it should not be afraid that its good work will place it in a difficult position in the following year. All the main levers of centralized planning — prices, finances, the budget, accounting, large capital investments — and, finally, the value, labor and major physical indices of rates and proportions in the sphere of production, distribution and consumption will be determined centrally.

Their fulfillment will be guaranteed and ensured by the fact that the economic councils (and the executive committees of the local soviets) will be presented with obligatory annual target figures for all the major indices. The economic council will become something more than merely a transmission point, which, unfortunately, it has often been up till now. It will be a center or junction at which all the threads of local planning converge. Its superior bodies will send it generalized state assignments for the economic area as a whole. The economic council will get from below the plans worked out by the enterprises themselves. These plans will take into account the obligatory assignments with respect to the quantity and assortment of output. In compiling their plans, the enterprises will be vitally interested in providing a maximum of output with minimum current and capital expenditures, and therefore one can assume with sufficient confidence that the sum of the plans of the enterprises subordinated to the economic council will fully ensure the fulfillment, and even the overfulfillment, of the assignments elaborated centrally.

The enterprises will not be hostile to the proposals of the economic councils concerning the improvement of certain indices, since the incentive base will not change in the least: the scale of allowances will remain the same. With the improvement of the plan, it will merely provide a greater reward for the enterprise. At present, every alteration of the plan (and there are dozens of these in a year) results in a demand by the enterprises for revision of all the related indices.

Many economists believe that enterprises can be stimulated to find reserves through improvement of the accounting base, a centralized system of technical norms, etc. There is no doubt that calculations and norms are necessary. However, these should be arrived at by the enterprises themselves. The enterprises should have a material interest in ensuring that these norms are progressive.

Thus, the system proposed proceeds from this principle: what is profitable for society should be profitable for every enterprise. And, on the contrary, what is unprofitable for society should be extremely unprofitable for the staff of any enterprise.

Some economists say that we must not place too much emphasis on profit, that it is a capitalist index. That is wrong! Our profit has nothing in common with capitalist profit. The essence of such categories as profit, price, and money is quite different with us, and they successfully serve the cause of the building of communism. Our profit, given planned prices on products of labor and the use of net income for the good of all society, is the result and, at the same time, a gauge (in monetary form) of the actual efficiency of labor outlays.

There are apprehensions that the enterprises will start putting out only profitable goods, rejecting unprofitable output. But, first of all, the violation of assortment will deprive the enterprise of any incentive bonus. Second, it is not at all beneficial that certain types of our goods are highly profitable, as a result of the shortcomings of price formation, while others are unprofitable. It is actually the task of price formation, as one can see from the decisions of the 22nd CPSU Congress, to ensure, under normal conditions, profitability for the whole assortment of goods.

It is also feared that the enterprises will begin to raise artificially the prices on new goods. But under present conditions it is actually hardest of all to control this from the center or even from the economic councils. The fact is that consumers are fairly indifferent to the calculations of the suppliers: they need the goods they want at any price, as long as the price was approved somewhere. The system proposed by us will change the situation. Any overstatement of prices on the goods supplied will reduce the profitability of the consumer. And this means that they will be as exacting as possible in checking the prices set by the suppliers. This will help the economic councils and the state planning committees to exert actual, and not formal, control over price formation.

At present, profitability falls if enterprises master the production of many new goods or new machinery. That is why we have worked out a scale of additional bonuses and reductions in bonuses that depends on the share of new goods in the plan. Given the manufacture of goods long in production, the size of the bonuses will be somewhat reduced; with the manufacture of new goods, it will grow substantially.

Moreover, the very process of price formation should be a flexible one. The prices on new goods that are more effective in production or consumption should be fixed initially so that the producing enterprise can cover its additional expenditures. This will do no harm to the consuming enterprise; on the contrary, it will derive profit for itself and the national economy. Thus, the reward for profitability can become a flexible tool in the struggle for the earliest possible introduction of new machinery and for improvement of the quality (durability and reliability) of goods. The existing system for encouraging enterprises to reduce costs of production or to increase output, as against the plan or as against the previous year, is a direct obstacle to raising the quality of produce and mastering the manufacture of new items.

In an attempt to find a way out, economists look for new and better "indices." There is no doubt that many indices have to be corrected. But even "ideally" constructed indices cannot produce any results by themselves. The problem is not one of indices, but of the system of relations between the enterprise and the national economy, of the methods of planning, assessing, and encouraging the work of production collectives. We already have the example of a new index, the "normative cost of processing," being substituted for the gross output index in the clothing industry. This was done to give clothing factories an incentive to manufacture clothes out of cheap fabrics rather than expensive ones. And what happened? It turned out that an opposite effect was produced: the factories readily make clothes that have long been in production out of cheap fabrics, for which the "processing cost" norm has become profitable, but they are very reluctant to make clothes in new fashionable styles out of the more expensive fabrics. But this is actually what the population demands. The inference is that any indices will be distorted when they are imposed upon the enterprises from above by the allocation [razverstka] method.

Instead, the enterprises should be given the opportunity to determine for themselves the optimal combinations of indices so as to achieve the ultimate effect: the best output, really required by the consumers, with the greatest profitability of production. Without such "freedom of economic maneuvering" it is impossible to increase sharply the efficiency of production. Without it you can talk as much as you like about increasing the rights of the enterprises, but you will not achieve it.

That is why the current plan should be relieved of the function of a gauge for determining the degree of reward; this will enhance the significance of the plan as a regulator of production for the purpose of ensuring the growth of the volume of production and its efficiency.

Naturally, we are far from the thought that the method proposed is some sort of panacea that will eliminate all the shortcomings by itself. It is clear that the organizing, educational, and monitoring work of the party and economic apparatus will remain the decisive force. However, this force will increase many times if it is supported from below by a total interest in the success of the matter, not for the sake of "indices," but in the name of the actual efficiency of production. In the process, there will be a sharp reduction of the administrative apparatus.

We should note that the system proposed will compel the enterprises to put out only such goods that will find a market and be paid for. Moreover, enterprises will calculate properly the efficiency of machinery and will no longer thoughtlessly order any and all new equipment at the expense of the state.

The custom now is to consider that any assessment of the work of enterprises and any kind of reward for them should proceed from the fulfillment of the plan as the most reliable criterion. Why is this so? It is alleged that the plan creates equal conditions for enterprises, that it takes into account differences in natural conditions, and in the degree of mechanization, and other "individual circumstances." In reality, plans for enterprises are worked out at present according to the so-called "report base," i.e., proceeding from the level achieved. And this actually creates very unequal conditions — favorable for enterprises doing poor work and strenuous for those enterprises which really find and utilize their own reserves. Why should one strive for better work under such conditions? Is it not more simple to try and get a "good" plan? It is high time to correct such a system.

Is it not clear that genuinely "equal conditions" can be provided when there is a uniform norm of profitability for enterprises that are approximately in the same natural and technical conditions? It is less dangerous to ignore some of the differences in these objective conditions than to level out the quality of economic management. By such levelling out we retain the backward methods in production. Let the enterprises themselves, provided with a production program and a long-range norm of profitability from the center, show what they are capable of achieving in the emulation for the best results. It is true that we do not have any competition [konkurentsiia], but this does not at all mean that we have no emulation [sorevnovanie] for the best methods of management. On the contrary, such emulation should be given full scope in our country.

Thus, what concrete proposals are being made to improve the state of affairs?

1. It should be established that the plans of enterprises, after the output program covering volume and assortment is coordinated and approved, are to be compiled in their entirety by the enterprises themselves.

2. In order to guarantee an honest approach in matters of interest to the state and to interest the enterprises in the maximum efficiency of production, it is proposed to establish a single fund for all types of material incentive and to make it dependent upon profit (in percent of the production assets).

3. It is necessary to approve centrally the scales of reward, in the form of long-term norms, which will depend on the profitability of various branches and groups of enterprises that

are in approximately the same natural and technical conditions.

4. It is necessary to intensify and improve centralized planning by informing only the economic councils (executive committees, departments) of the obligatory assignments (target figures). It is necessary to do away with the practice whereby the economic councils impose assignments on the enterprises in accordance with the "level achieved." The economic councils should be ordered to check, assess and improve, on the basis of economic analysis, the plans worked out by the enterprises without changing the profitability scales as the basis for the encouragement of enterprises.

5. It is necessary to work out the procedure for utilizing the single incentive funds formed from the enterprises' profits, with a view to extending the rights of the enterprises in expending their assets for the needs of collective and personal encouragement.

6. It is necessary to establish the principle and procedure of flexible price formation for new goods so that the most effective goods will be profitable both for producers and consumers, i.e., for the national economy as a whole.

Kommunist, 1962, No. 18

The Role of Profit in a Socialist Economy

L. Gatovskii

This article examines the new content of profit in the socialist economy; the place of profit in the system of economic categories of socialism; shortcomings which restrict the possibilities of using the index of profitability; and ways of enhancing the role of profit as a means for improving the operation of socialist enterprises.

***

Problems of improving planned management of the economy, raising the efficiency of production, rationally using and saving resources, and mobilizing internal reserves have acquired primary significance in the light of the economic tasks put forward by the Program of the CPSU. The Central Committee of the CPSU, at its Plenary Meeting in November 1962, outlined concrete ways for accomplishing these tasks. Acceleration of technical progress, maximum development of new, highly promising branches, introduction of progressive production methods, elimination of primitive methods and routine from economic life, and utmost specialization and concentration of production — all this is aimed at stepping up the growth rates of labor productivity, reducing production costs, and raising the profitability of the economy.

Problems of the sources of accumulation and of increasing the mass of profit and its rational use for the successful accomplishment of the sweeping tasks of building the material and technical base of communism occupy an important place in the economic policy of the CPSU, in the development of the entire system of economic management. And this is fully natural.

Profit comprises a very substantial part of the entire net income of society. Together with the other part of the net in-

come, the turnover tax, it represents, in money terms, the
value of the surplus product created in all branches of material production. Profit is a vitally important source of expanded socialist reproduction. Under the 1963 plan, the entire
mass of profit in the Soviet national economy will reach the
huge sum of 35.7 billion rubles (11.3% more than in 1962),
exceeding the turnover tax — 33.8 billion rubles. More than two-
thirds of the profit in 1963, 26.1 billion rubles, will go into the
state budget for financing the economy, social and cultural requirements, the needs of state administration, and the country's
defenses. About one-third of all the profit of the state enterprises will remain at their disposal for developing production,
satisfying the material and cultural requirements of the collectives and individual workers, and paying bonuses.

Since profit reflects reduction in production costs, higher
labor productivity, savings of productive resources, improvements in production organization, growth of output, etc., it is one
of the necessary value indices of the economic activity of an enterprise, of the efficiency of its operation. The magnitude of
profit and the rate of profitability (ratio of profit to the cost of
production) enable the state, to a large extent, to evaluate the
operation of a given enterprise.

Since profit is an index of the economic results of an enterprise's operation, it functions as one of the economic stimuli to
improving its operation. This is connected with the entire system of cost accounting and assessment of the activity of an enterprise, with the use of profit as a material incentive and the
corresponding organization of a bonus system. Profit is the
source for making up such an incentive fund as the enterprise
fund. Interest in profitability prompts an enterprise to improve
its operation.

Higher profitability is one of the economic criteria of relative efficiency of production for different kinds of goods which
satisfy a certain requirement of society. What is needed is that
greater profitability should really reflect lower production
costs, smaller expenditures of labor per unit of output, and
that priority development of more profitable goods should conform to the interests and needs of society. Recoupment periods
and, consequently, the level of profitability serve as important
guideposts in choosing variants of capital investment for the
planned output of particular goods.

Thus, the objective necessity of profit in the socialist
economy is dictated by a number of factors. First, socialism

develops on the basis of expanded reproduction and, consequently, it is impossible without the production of surplus
product. Second, socialist reproduction inevitably takes place
in the value form as well; therefore the socialist surplus product appears in its monetary expression, i.e., as the net income
of society. Third, the net income of society necessarily must
assume the form which serves society as an index of the efficiency of operation of the basic link of the national economy,
namely, the enterprise, and as a stimulus to its improvement.
The category of profit or the net income of an enterprise constitutes such a form of the net income of society.

In this connection, we must briefly examine the distinctions between two categories of one type which make up the
bulk of the net income of society — between profit and the turnover tax. The turnover tax is a specific form of mobilizing the
centralized net income of the state. The turnover tax has some
advantages from the standpoint of providing a guaranteed and
swift influx of money into this centralized fund. The rates of the
turnover tax are fixed in advance for each unit of output. The
machinery for collecting the turnover tax also exercises financial control to a certain extent. But, in contrast to profit, the
turnover tax does not serve, for an enterprise, as an index of
profitability, a criterion of its work, a stimulus and source of
material incentive. The rates of the turnover tax per unit of
output are invariable, no matter how the cost of production of
this unit changes; increases in this rate do not depend on the
enterprise. The situation is different as regards profit per unit
of output: its size depends on the efficiency of operation of the
enterprise, since, by reducing its production costs, it increases
profit.

That is why, as distinct from the turnover tax, profit is
a cost-accounting category which exerts a big stimulating influence on an enterprise. It is accumulated by the enterprise itself, and a substantial part is used for its own needs. Profit is
an integral link of the mechanism of cost accounting. The essence of cost accounting is that any enterprise should cover its
expenditures with its own income and should have a profit over
and above this. The system of cost accounting makes every enterprise interested in obtaining a bigger profit and, consequently, in making better use of labor, material, and financial
resources. We can imagine, in the future socialist net income
without such a form as the turnover tax (although at present it
is necessary), but without profit, without a system of cost

accounting, normal economic management in a socialist society is simply inconceivable.

Profit is a very important element in the planned utilization of the law of value, requiring that prices at which an enterprise sells its output should be based on the socially necessary outlays. Thus, the higher the individual outlays of an enterprise as compared with the socially necessary ones, the worse the results of its operation in terms of value; and the lower the individual outlays, the better these results. It is here that profit appears on the scene as the concrete form through which the operation of the law of value impels an enterprise to cut its individual outlays, to save living and materialized labor. The relationship between individual and socially necessary outlays is actually manifested in the profitable or unprofitable operation of an enterprise; profit stimulates an improvement of this relationship.

The category of profit is inseparably bound up with the law of socialist accumulation. In economic literature its role is usually limited to the sphere of distribution of the net income. Yet it is above all the production of net income that is of primary importance for the development of our economy. From this follows a major requirement of this law: each socialist enterprise, as a rule, must produce net income, must be profitable. Such is the normal condition of socialist reproduction. The economic law of socialist accumulation contains three organically interconnected elements: the need for the production of net income, the necessity of a distribution which ensures expanded reproduction at the required rate, and, lastly, the subordination of the process of production and distribution of the net income to the aim of advancing the material and cultural standards of the people. This reveals the deep antithesis between socialist and capitalist profit.

In bourgeois society profit is an outcome of capital. It represents a form of surplus value, and reflects the very essence of the exploitation of labor by capital. Under socialism profit is a form of the socialist net income of society which is produced by labor free from exploitation and is distributed in the interest of the people. The systematic growth of profit subordinated to this aim is the source of the continuous expansion of socialist production on the basis of the latest technology. The productive use of the growing mass of profit ensures, economically, technical progress and growth in the productivity of social labor. In turn, operation of the law of the steady growth of labor productivity creates the possibility for systematically increasing the mass of profit. The formation of profit and its distribution and use are connected with the entire process of social reproduction. That is why profit is one of the main objects of state planning and serves as one of its economic instruments for developing socialist production and mobilizing the internal reserves of our enterprises.

While in bourgeois society the growth of profit is an aim in itself which the capitalists seek to achieve by any means, and try to overstep any boundaries here, the increase of profit under socialism is one of the means for the achievement of the aim of socialist production — to satisfy most fully the requirements of the people. It is used to the extent to which it facilitates the achievement of this aim. The aim is the criterion which determines the bounds for using profit. The principle of obtaining maximum profit at any price, irrespective of how this affects the interests of society, the consumers, working and living conditions, etc., is deeply alien to socialism. It is no accident, for example, that under socialism the improvement of working conditions is an independently important reason for introducing new technology. Profit is one of the subordinated elements in the entire system of the economic categories of socialism.

The analysis of the nature of profit under socialism given by Comrade N. S. Khrushchev in his report at the Plenary Meeting of the Central Committee of the CPSU in November 1962 is of great scientific and practical significance. He stressed that "in characterizing the socialist system of economy, we must not confuse the concept of profit as applied to the entire national economy and as applied to a particular enterprise." If we take the socialist system of economy as a whole, profit does not have the social meaning which characterizes it in capitalist conditions. In the capitalist economy profit is the aim of production. But in the socialist economy, where the aim is to satisfy the requirements of society, goods are produced not to obtain a profit, but because they are needed by all of society. "An individual enterprise, however, is a different matter," N. S. Khrushchev said. "In the given case, the question of profit is of great importance as an economic index of the efficiency of its operation. How an enterprise works — at a loss or at a profit, whether it eats up social resources or multiplies them — is of great importance. Without an account of profit it is impossible to determine at what level an enterprise operates and what contribution it is making to the fund of the entire people."

The antithetical aims of production under capitalism and socialism also imply a fundamental difference in the role which profit plays in the distribution of the means of production and labor. In bourgeois society capital moves to branches where profit is higher. In the socialist planned economy the resources are distributed between branches so as to satisfy the totality of the growing requirements of society, in line, of course, with the available production potentialities. Moreover, in socialist conditions it would be absurd to take differences in the level of profit as grounds for giving preference to some basic requirements at the expense of others.

The aim of socialist production is concretely expressed in the national economic plans. They set a definite level of satisfying the requirements of society and the attendant volume and structure of social production, which can be determined by proceeding from the scale of the economy as a whole, and not from the position of separate enterprises and branches. Here is a manifestation of the difference in the very concept of profit and its role as applied to an enterprise and to the national economy as a whole. It is one thing to plan the national economy: here we deal with the basic proportions in the economy of a country. One cannot correctly determine these proportions according to the principle of channeling investments where profit is higher. Other questions arise when we deal with the operation of an individual enterprise. The national economic plan and the proportions established in it determine the tasks of the given enterprise. The question now is how to utilize its internal resources for best accomplishing these tasks. Here we cannot get along without the economic criterion of profit. It does much to indicate how successfully the enterprise is fulfilling its mission in the socialist system of economy, what the balance sheet is of its economic relations with all of society, of which it is a small part, how efficiently it is managed, how efficient is the production of each type of goods. In this case other, specific conditions for the wide application of the category of profit arise.

Consequently, the role of profit becomes different when we move from the scale of the national economy as a whole to branches and, particularly, to enterprises. Let us take the question of profit as a criterion for choosing where to channel resources for the development of separate categories of produc-

tion. After the level and structure of satisfying the requirements of society have been determined, the question arises
of how to satisfy them best with the least outlays. What category of production is economically more efficient, cheaper,
and more profitable? Where, consequently, is it more advantageous to direct the resources for satisfying the given requirement? For example, which of the interchangeable kinds of
output should be developed? Should we produce fabrics from
natural or synthetic raw materials, pipes from metal or from
plastics? What building materials should we use for carrying
out the housing construction plan, and so on? The answers will
be furnished by calculations of comparative economic effectiveness. It is by no means a matter of indifference to society
in which of the interchangeable kinds of output the invested
resources will be recouped faster, where production costs will
be lower and profit higher. This applies to the choice of the
direction of capital investments for the production of interchangeable goods in various branches and the same kind of
goods within one branch: at what enterprises, with what equipment and technological processes primarily to develop the
given category of production, how to allocate the capital investments, etc. Here profitability is one of the criteria.

The effort to raise profitability exerts a great stimulating
effect on an enterprise. Its essence lies in systematic improvement of the organization of production, increasing the technical
facilities per worker and labor productivity, and in economies
and rational utilization of production resources.

All this shows how wrong is the position of the economists
who claim that the average rate of profit in the national economy
should be the single standard of efficiency of production, serv-
ing as the basis for allocating resources. Under capitalism,
the spontaneous movement of capital inevitably gives rise to
the tendency toward an average rate of profit. For us, however,
this is an unsuitable scheme, which runs counter to the essence
of the socialist economy. Some economists, correctly wishing
to elevate the role of profit in our economy, wrongly regard
the growth of profitability as some kind of automatic regulator
which supposedly is capable, by itself, of directing the operation of an enterprise into the channel required by society. But
experience constantly demonstrates that the level of profitability
can in no way serve for us as a single absolute criterion of the
success of an enterprise and replace the other value indices of
production, not to mention indices in physical terms. We are

talking about indices of production of use values (volume,
structure, quality of output), the introduction of new articles
and new technology, rational use of living and materialized
labor, etc.

***

An obvious belittling and, at times, outright ignoring of
the importance of profit and of value categories in general
were characteristic of the period of the cult of Stalin's personality. He substituted naked administration by fiat for economic
instruments of directing the economy and material incentives.
He implanted subjectivism in planning and economic management. At times this subjectivism assumed the nature of willful
arbitrariness. His verbal declarations about the need for taking
full account of the economic laws of socialism were at variance
with practice, which often ran counter to these laws and economic categories. All this had a direct effect on the attitude
toward profit, which was regarded as a purely formal category.
The objective significance of profit consists, above all, in the
fact that it is a necessary part of the price. But at that time the
approach to price itself was voluntaristic. Price was artificially
divorced from its objective basis — value. It is well known
that the operation of the law of value in the socialist economy
was generally denied for a long time with the blessing of Stalin.
It was claimed that price exists without value. Such a concept
of price was far removed from the Marxist materialist understanding of price as an economic category, ultimately expressing the outlays of socially necessary labor. In fact, this was
an idealist understanding of price as an object of the arbitrary
action of state agencies.

In 1941, in a conversation with economists, Stalin declared
that the law of value existed in the Soviet Union. But the very
restricted interpretation he gave to this law and to value categories, including price, in effect changed little as compared
with the former concepts. The law of value was so "curtailed"
that, actually, its role was denied. The categories of commodity
and value were declared to be incompatible with state ownership of the means of production, and were not "admitted" into
the sphere of production of the means of production. Of these
"old categories of capitalism," as Stalin put it, only the "outward aspect" remained.

Stalin recognized the law of value as a law of production
only for collective farm production. But actually even this was
reduced to naught by his assertion that planned procurement

prices of agricultural produce, i.e., the prices at which the
collective farms sold the bulk of their production for the market, could not be based on value. He declared that value was
completely incompatible with planned price formation. And so,
according to Stalin's propositions, the principle of "price without value" was retained both for the state and the collective
farm sectors. Wide scope for subjectivism was opened in the
practice of price formation. Prices were often constructed without regard for the socially necessary outlays, and this was
sanctified by corresponding dogmas.

In these conditions, the question of profit as a necessary
element of price lost any foundation whatever. If, for example,
wholesale and procurement prices were far below the cost of
production, and all this was justified "theoretically," what
sense was there in raising the question of the objective necessity of profit! No wonder that in that period profit could not
be regarded as a necessary category of a normally operating
socialist enterprise. Since prices of collective farm produce
were much below the cost of production, there were economists
who "eliminated the problem" by declaring that the categories
of profitability, cost accounting, and production costs were
generally inapplicable to the collective farms. Attempts to
take into account the element of profitability in choosing the
objects of capital investments even within one branch, or for
interchangeable kinds of goods, and also in picking the areas
for the location of enterprises, were often characterized as
bourgeois methods.

Any attempts to compare profit with the productive assets
were regarded in the same way. This attempt as such was regarded as transferring the capitalist categories of price of
production and average profit to our soil. Things went to such
lengths that, in general, operation at a loss was regarded
indiscriminately as a normal economic phenomenon, and even
as a manifestation of the advantages of socialism. The opinion
was widespread that the principle of profitability in general is
a capitalist principle, alien to socialism: it, you see, restricts
the advantages of a planned economy and therefore must be
discarded.

In his Economic Problems of Socialism in the USSR
[Ekonomicheskie problemy sotsializma v SSSR], Stalin counterposed to the principle of profitability of enterprises his own
principle of higher [vysshaia] profitability; he included the
latter among the advantages of a planned economy, which ensure

the uninterrupted growth of production. Moreover, he regarded profitability of enterprises as "temporary and unstable," which "in no way can be compared with this higher form of stable and constant profitability." Here, in the first place, the concept of higher profitability was divorced by Stalin from profit. Therefore, the very term "higher profitability" does not conform to its content. The clear and definite concept of profitability was replaced by the absolutely hazy, indefinite, and essentially meaningless concept of higher profitability, which had no relation to profit. Second, the attitude here toward profitability of enterprises was wrong. Why can't it be stable and permanent? Why is it doomed to be "temporary and unstable"? Third, the planning which Stalin associated with "higher profitability" was counterposed to the profitability of enterprises. This break between planning and profitability runs counter to the fundamental tasks of developing the socialist economy. One of the primary tasks of planning is to ensure profitability of individual enterprises as the basis of socialist accumulation. Whereas capitalists are directly interested only in the profits of their enterprise and not the total profit in the country's entire economy, the socialist state sees to it that the total profit of society rises on the basis of the growth of profit at each enterprise. Why shouldn't the advantages of a planned economy affect the growth of profitability of enterprises? After all, the planned economy opens up extensive sources of accumulation formed from the profits produced at enterprises. At the same time, the continuous expansion of production is impossible without accumulation, the normal requisite for which is the profitability of enterprises. The concept of "higher profitability" undermined the significance of profit and objectively led to justifying operation at a loss by many enterprises. The line of reasoning was as follows: of what importance is the unprofitable operation of an individual enterprise if we have a planned economy and higher profitability!

Recent years have witnessed a sharp change in the attitude toward profit both in economic theory and in practical economic activity. The Party has resolutely cast aside the subjectivist propositions which gained currency during the period of the cult of Stalin's personality, including the grossly mistaken approach to such categories inherited by socialism from the past as commodity, value, price, and profit. They began to be regarded in their new, socialist quality, and not as outward, superficial forms, not as instruments for willful planning which could be used entirely arbitrarily to fill any economic content, to establish any quantitative and qualitative relationship, but as objectively necessary categories inherent in the socialist economy. The significance of profit as an indicator of the operation of an enterprise has been enhanced. The process of increasing profit and reducing the number of unprofitable enterprises has been stepped up. The stimulating role of profit for enterprises has been enhanced, and they have gained greater possibilities for using and disposing of profit.

The measures carried out in price formation are of great importance for raising the role of profit. In a number of industries the level of prices has been brought close to value, thereby providing great possibilities for enterprises to achieve profit by their efficient work. Reduction of production costs also had this effect by helping to eliminate operation at a loss. The changes in this respect can be confirmed by the rise in the level of prices of agricultural produce. This process, designed to ensure the profitability of collective farm and state farm production, began in 1953 with a sharp rise in procurement prices, and continued up to 1962, when the purchase prices of animal products were raised. Much has been done to enhance the role of profit as an effective stimulus to the development of socialist production. But in the economic activities of our enterprises profit has not yet assumed a place corresponding to the tasks of communist construction. The power of inertia is still making itself felt in an underestimation of the category of profit both in economic science and practice. The Party is taking measures to correct this situation. In his report on the Program of the CPSU at the 22nd Congress, N. S. Khrushchev stressed: "We must elevate the importance of profit and profitability. In order to have enterprises fulfill their plans better, they should be given more opportunities to dispose of their profits, to use them more extensively to encourage the good work of their personnel and to extend production."

Further adjustment of prices is a prerequisite for elevating the role and significance of profit. Profit is formed directly from the difference between the price and cost of production. That is why its growth may be caused both by a reduction of production costs and by high prices. The socialist approach to profit is that an enterprise should work for higher profitability by reducing production costs, saving living and materialized labor, and improving management. If a price is set too high

and automatically ensures the enterprise a big profit, this
undermines the stimuli to cut production costs, raise labor productivity, and use resources rationally. An "easy life," hot-
house conditions, are created for an enterprise which enable it
to be highly profitable without any exertion of effort. The general line in price formation is an economically justified reduction of prices. It must correspond to real conditions of production and enable the enterprise, by its efficient operation, not
only to cover outlays, but also to obtain a profit. This means to
economically induce enterprises to mobilize internal production
reserves and thereby reduce production costs so as to have,
even when prices are cut, a sufficiently high profit.

Experience has shown that we need not only periodic general revisions of prices, but also systematic current measures
for adjusting prices in particular industries or sectors of the
economy. If prices are not revised for a long time a big dis-
parity in profits will inevitably arise, placing some enterprises
in an undeservedly privileged position and artificially increasing
their profits. As a result of changed conditions of production,
often for reasons unrelated to the efficient operation of enterprises, but, say, due to radical technical reconstruction, the cost
of production of particular goods is sharply cut. Inasmuch as
prices are not changed, profit reaches huge proportions, as
much as 50 to 100% at times, while other enterprises, notwithstanding their efficient operation, have an entirely different ratio
of prices and production costs.

It is not only an inflated price and too high a profit, but also
an excessively low price which artificially cuts profit or makes
operation unprofitable, that weakens the stimuli to an improvement in work. Enterprises are thereby deprived of the prospect
of becoming profitable. Both inflated and reduced prices adversely affect material incentives and the system of cost accounting. Excessively low prices are one of the reasons for the
substantial number of unprofitable industrial enterprises. Although their share has declined greatly, even now about 20% of
all industrial enterprises under the jurisdiction of economic
councils are operating at a planned loss. In a number of cases
the plan is based in advance on the premise that, at the given
level of prices, enterprises cannot be profitable. There still are
wholesale prices set considerably below the cost of production,
which predetermine the unprofitable operation of entire
branches of industry.

The extent to which profit can discharge its role as an

index for assessing the efficiency of an enterprise and stimulating it accordingly largely depends on the regulation of prices,
on how precisely prices reflect the socially necessary outlays,
on the relationship between the price and the normal costs of
production for the given enterprise. The unjustified diversity
of profitability in producing various kinds of goods, a diversity
associated with prices, is still very great. According to available data, profitability in the machine-building industry ranges
from 5 to 60%. In machine-building plants of the Moscow city
and regional economic councils, the level of profitability fluctuated from 4.4 to 56.5% in 1961, and at Leningrad machine-
building plants, from 10 to 122%. Some plants fulfill their profit
plans through highly profitable secondary items, on which profit
reaches 100% and more, while their main goods are produced
at a loss. This diversity in profitability actually conceals real
differences in the efficiency of operation of enterprises.

Particularly adverse consequences result from an improper ratio between prices and profits for obsolete goods, on
the one hand, and new, progressive articles, on the other. The
existing procedure for setting prices frequently does not give
enterprises an incentive for developing new technology, new
and improved goods. In the case of machines whose output was
mastered long ago, the reduction of production costs over a prolonged period usually sends up the rate of profitability much
higher than 5%, and often up to 20 to 30%. Yet, frequently, for
new articles, new machines, the manufacture of which is being
mastered, prices are set which provide for a profit rate not
higher than 5%, and at times even lower. Let us take Moscow
enterprises as an example. A number of plants producing a
stable output which has remained unchanged for years had
profits of 35 to 45% in 1961. On the other hand, at enterprises
in which the output is swiftly changed (for example, plants
producing special and multi-purpose machine tools and automatic transfer lines), the rate of profitability is 4 to 5%. Thus,
the price and, thereby, also profit very often make the manufacture of new technology disadvantageous, especially if account is taken of the additional outlays, often unforeseen when
gearing production to new models, the additional trouble, production risk, etc., and, on the contrary, too high a profit prompts
enterprises to continue putting out obsolete goods.

Thus, improper price fixing not only nullifies the stimulating role of profit, but also creates the possibility of its having
an adverse influence on production. It gives rise to a desire on

the part of enterprises, alien to the principles of socialist management, to raise their profitability not through better work, but through narrow commercial methods, the quest for an advantageous "market situation" through the choice of an assortment of goods which gives a high profit, although, as often happens, it is less needed by society and yields a smaller effect. The significance of profit as an index of the efficiency of production is weakened and often undermined. Frequently, owing to the price factor, the profitability of enterprises rises while their indices of production costs and labor productivity deteriorate, or profitability drops while these indices improve.

In his report to the Plenary Meeting of the Central Committee of the CPSU in November 1962, N. S. Khrushchev pointed to the abnormal situation as regards prices of industrial goods, which impedes the elimination of many serious shortcomings in planning production, in applying cost accounting to the full, and in providing conditions for the profitable operation of enterprises. The shortcoming in price formation can be described in the most general way as too big a deviation above or below the socially necessary outlays, which results, regardless of the operation of the enterprise, either in an unjustifiably low profitability, and even loss, or in excessive profitability.

**

A substantial revision of wholesale prices in industry is being prepared at present. Within the bounds of the present average level of prices, they will be substantially leveled on the basis of the cost of production so as to ensure more normal profits. This revision of prices will substantially improve the situation in price formation and will be an important step toward implementing the directive of the CPSU Program that "prices must, to a growing extent, reflect the socially necessary outlays of labor and ensure a return of production and circulation expenditures and a certain profit for each normally operating enterprise."

But this revision of wholesale prices by itself, of course, cannot solve all the problems involved in improving price formation. Thus, for example, there is an urgent need to improve the zonal system of setting agricultural prices, to ensure more justified and expedient differentiation by zones. A number of problems remain to be solved in improving industrial prices. We pointed earlier to the need for systematic, current revision of prices to remove the inevitably accumulating disparities between prices and production costs, and, thereby, for greater leveling out of the conditions of profitability of enterprises. It is particularly important to establish more correct ratios between the prices and, therefore, profits for obsolete types of output and those for new, progressive kinds of goods. Moreover, it is necessary to make not only the producing enterprises, but also the consuming enterprises highly interested in new kinds of goods, in new technology. There have been a number of proposals on adapting prices more to the requirements of technical progress. Specifically, recommendations to fix prices depending on the "age" of the goods merit attention. Prices and, consequently, profit should be more highly differentiated depending on the quality and grade of goods. We ought to study methods proposed for radically eliminating the unprofitable operation, artificially created by flaws in prices, of many industrial enterprises. Even after the forthcoming revision of wholesale prices in industry, the number of unprofitable enterprises, though sharply cut, will most likely remain considerable.

Every possible measure should be taken to reduce to naught, to restrict to special exceptions, the so-called planned operation at a loss, which spreads sentiments of dependence, strikes at cost accounting, and stifles the initiative of enterprises. To begin with, we should intensify the drive for improved organization of production; against unproductive outlays, which often grow not only in absolute figures but also relatively; against mismanagement. But we cannot forget the price factor. It often happens that, at the given average price of an industrial item, some plants get a profit while others operate at a loss, and not because the former are more efficient than the latter, but because, regardless of their work, some have much better production conditions than others, thus making the average wholesale price unprofitable for the latter. We ought to study the proposal that, in such cases, wider use be made in some branches of the system of setting lower and higher factory prices while preserving one price for consumers. This would help eliminate unprofitable operation for which the enterprise is not to blame.

Thus, as price formation is improved, the role and significance of profit as an index of the efficiency of operation of an enterprise, of the living and materialized labor it saves, will continue to rise. The state will then be able to use profit more widely and effectively for stimulating enterprises in the interest of society.

At the same time, adjustment of prices alone is not enough to improve the profit index to the utmost. The policy of approximating prices with socially necessary outlays does not at all mean that they must fully coincide. Price, in addition to the accounting function, also has distributive and stimulating functions which necessarily demand that it deviate from value somewhat. Profitability will continue to depend, though to a lesser extent, on changes in the structure of production; in other words, profit will grow not only through factors which depend on the efficiency of operation, but also on more advantageous prices, on changes in the structure of output. How can we exclude the influence of the price factor, "clear" the profit index of it in assessing the operation of an enterprise and stimulating it? This can be done by taking into account the influence of factors associated with changes in the structure of output. Further, to exclude other factors which the enterprise has no control over, we must proceed, in evaluating the work of an enterprise, not only from the profit shown in the balance sheet, but also from the profit made on the output in some cases, and on the sale of the output in other cases.

What are the main directions along which the further elevation of the role of profit is proceeding and will proceed? First, the growth of the mass of profit is being accelerated, which raises the importance of profit as a source of expanded reproduction. This is achieved both by a general increase in the entire surplus product, and by a rise in the share of profit in the total net income of society while the share of the turnover tax is falling. The total sum of the turnover tax in 1960 rose by less than one-third as compared with 1953, while the sum of profit nearly trebled. If we take the last four years (1962 as compared with 1958), profit rose by 64% and the turnover tax by only 6%. That profit grows much faster than the turnover tax is a progressive tendency, reflecting the policy of further consolidating cost accounting. This tendency must be continued, taking full account of state interests.

Second, in accordance with the policy set at the 22nd Congress of the CPSU and the Plenary Meeting of the Central Committee of the CPSU in November 1962, of extending the economic rights of enterprises, it is necessary to considerably increase the share of the total sum of profit which remains at the disposal of enterprises and is used by them to expand and improve production, increase their assets, replenish their circulating funds, and for bonuses and the social and cultural needs

of their personnel. Regulation of the use of financial resources
by enterprises, where it is excessive and too detailed, should
be eliminated, and enterprises should be given greater opportunity to maneuver with these resources. This presupposes
extension of the rights of enterprises in the sphere of financial
planning and expenditure of resources. Then enterprises will
be able to utilize the financial results of their successful operation more effectively. This will really enhance their interest in
the profitability of production. Thus, there will be greater encouragement of high efficiency of operation through the profit
index. In recent years — 1962 as compared with 1957 — the
sum of profit remaining in the economy has doubled. This
process should be accelerated, with the simultaneous strengthening of the centralized bases of planning and transmitting the
major plan targets for enterprises, including the targets for
labor, wages, accumulation, and capital investment.

Third, we must see to it that, through a reduction of costs
and a set of measures for reinforcing cost accounting and
further regulating prices, each well-operated enterprise has
sufficient profit. This will ensure an ever greater evening up
of operating conditions, as a result of which the real importance of profit as an economic stimulus and more precise index
of efficient operation will rise substantially. In this connection,
it is of special importance to ensure the maximum stability
of plans and, when they do change, to take this into account
when assessing the work of the enterprises and providing them
with incentives. There are many instances in which plan assignments of enterprises are repeatedly changed in the course
of the year and run counter to their specialization and productive possibilities. It is clear how adversely this affects the
profitability of production. An end should be put to frequent and
unjustified changes in the profit plans of economic councils
and enterprises. In 1961, for example, the profit plan of the
Estonian Economic Council was changed 16 times.

Fourth, the role of profit as a source of funds for material
stimulation should be further raised. This, of course, does not
preclude but presupposes the payment of bonuses on the basis
of other indices as well. At the same time, after an appropriate
experimental test, it is advisable to gradually extend the practice of using profit as a source of funds for material stimulation and as an index for the payment of bonuses (including engineering and technical personnel). For this purpose we should
first of all take a branch in which profit can more exactly reflect

the efficiency of operation of an enterprise and where other indices do not occupy a special, specific place (for example, the growth of production in the extractive industries). From this standpoint the profit index can even now be given a preferential position in a number of branches of the manufacturing industry with a more stable nomenclature of goods and a prevalence of a comparable structure of output.

To prevent enterprises from being interested in getting reduced profit assignments, enterprises should be given material incentives not so much for overfulfillment as for fulfillment of the profit plan. The magnitude of the incentive should depend on the level of the planned assignment, the effort it involves. To make the fund of an enterprise more effective, allotments to this fund should take into account the average amount of profit per worker. At present the bonus funds do not play a sufficiently weighty role. Too many obstacles are often set up to the payment of bonuses. It is important to extend the possibilities for the justified award of bonuses. Of great importance would be a big increase in the ability of enterprises to pay bonuses in the shops for a saving of material outlays, to restore bonuses for a reduction of non-productive expenses, for a reduction of losses from spoilage, for rational use of materials and waste products.

The index of the ratio of profit to productive assets, to fixed and circulating assets (in other cases, for example, the index of the ratio of output volume to assets) should be used in some instances as an important method of encouraging the efficient use of productive assets. Moreover, an enterprise should be made interested not only in better use of assets, but also in their expansion, in the introduction of advanced technology, and in making such advance preparations in production which, without yielding an immediate effect, would ensure the success of future operations.

But however perfect the prices are, however improved and "cleared" the profitability index is, it cannot serve as the sole criterion for fixing the size of the incentive fund of enterprises and paying bonuses to the personnel. A differentiated approach to branches and groups of enterprises, consideration of their specific features, is needed. The strength of a planned economy, its advantage, lies specifically in the ability of society consciously to take into account all the diverse conditions and tasks confronting various industries and enterprises, to create a differentiated system of indices and incentives, to bring some

of them to the foreground in one situation and others in a
different situation, depending on the concrete conditions and
tasks. We now have a number of studies which provide economic grounds for the expediency of applying some or other
indices in assessing the operation of enterprises and stimulating them, depending on the peculiarities of branches and enterprises. The task now is to draw general conclusions from them
for practical application.

The proper employment of profit, with a full account of
the nature and advantages of the socialist system of economy,
is a prerequisite for the successful building of the communist
economy, for raising the efficiency of production, accelerating
its growth rates, and, consequently, for improving the living
standard of the people.

Pravda, September 21, 1962

The Plan Target and Material Incentive

(Concerning the Proposals of Comrade Liberman)

V. S. Nemchinov

Planning and material incentive are two most important factors in the development of the socialist economy. For the first time in the entire history of the development of the productive forces of society, social ownership of the means of production, constituting the basis of socialist relationships in production, eliminates the antagonism between the interests of the individual enterprise and those of society. The socialist mode of production unifies hundreds of thousands of enterprises into a structured system, operating and directed in accordance with a single national economic plan that has been discussed by the people as a whole and has the force of law.

Thus, planning is an inseparable component of the socialist economy. With respect to material incentive, however, the situation is somewhat different. The role played by material incentive with respect to growth of production has differed at various stages in the development of our economy. During the period when the economic chaos inherited by the Soviet state from the old system was being overcome, in a period when the question of "who will win over whom" was being resolved in the economy, when it was necessary to establish our own industry and a collectivized agriculture in a historically brief period, the interests of society as a whole often did not permit complete consideration of the interests of the individual enterprise. The years of World War II and the period of rebuilding the

---

The late Academician Nemchinov was Chairman of the USSR Academy of Sciences' Scientific Council on the Interdisciplinary Problem of the Scientific Foundations of Planning and the Organization of Social Production.

economy that had suffered from the fascist attack also required the concentration of all the strength of the Soviet people.

However, the concentration of the economic efforts of the Soviet state as a result of these circumstances never meant abandonment of the principle of material incentive. As we know, Lenin repeatedly emphasized the immense role of material incentive in the building of socialism. And current economic experience constantly reminds us of this. Socialist social production has attained a level of development at which all the available economic resources and hidden possibilities cannot be mobilized adequately and effectively utilized without an improved system of material stimuli.

In the past year, the economic literature has repeatedly published the statements of managerial, planning, and scientific personnel on matters of improving the management of production.

What is responsible for the appearance of these articles? The answer is that fulfillment of the vast program set forth by our party for establishing the material and technical base of communism and creation of the world's most powerful industry are possible only if there is continuous improvement of the forms of management of social production and a constant search for new economic techniques for controlling it.

The practical experience of recent years indicates that the plan targets for a large number of indices by no means always stimulate growth of production and assure mobilization of unutilized resources and the introduction of new technology. Planning that is based on the level of gross output, the cost of production, and the labor productivity already achieved by an enterprise inevitably leads to an effort on the part of managerial personnel not to reveal all their productive capacities and resources.

Naturally, such a situation begins to inhibit the operation of the irrevocable law of economic activity formulated in the Program of the Communist Party — to assure maximum results with minimum expenditures, inasmuch as part of the economic resources remains unutilized.

As a consequence, a broad discussion is now unfolding with respect to the most important economic problem of our contemporary economic science: how to improve the planning system so that enterprises and their personnel will have a material interest in increasing output, cutting production costs, and introducing new, highly productive equipment, so that they will have an interest in higher targets under the plan.

Comrade Liberman's recommendations are based upon introducing long-term norms into planning. This proposal is certainly of interest, for the establishment of such norms should serve to eliminate the harmful striving of heads of enterprises to be given understated plans, and will at the same time directly stimulate a more intensive development of production. That is why the experiment of the Economics Laboratory of the Kharkov Economic Council, whose data provided the basis for Comrade Liberman's article, has attracted the attention of the USSR Academy of Sciences' Scientific Council on the Interdisciplinary Problem of the Scientific Foundations of Planning and the Organization of Social Production. A report by Comrade Liberman, the head of the laboratory, was discussed at a plenary session of the council in April 1962. The council gave its support to the initiative of the man from Kharkov and recommended greater experimentation in this direction, turning its special attention to the elaboration of proposals to stimulate the production of new types of output and introduction of the achievements of technological progress.

It is necessary to note that the planning of the final rather than intermediate results of economic activity is acquiring increasing significance at the present time. Objections to the use of gross output as a planning index are being raised more and more often by managerial personnel. This index is a poor reflection of the results of the economic activity of the individual production team because it includes the value of raw materials, fuel, and other materials received from the outside. When planning is based on this index, it is found to be advantageous to utilize more expensive materials and services. By consuming more expensive materials and employing more costly services, one can overfulfill the plan for gross output. As a result increases in gross output are derived from the intermediate product. On the whole, however, society is interested in the growth of the final product rather than of the intermediate product.

From the viewpoint of the national economy, the final product is characterized by the material composition of the newly produced national income. For the individual enterprise, however, the final product is the output sold and shipped. However, the volume of net income created in the sphere of material production remains the general and sole index of the ultimate economic result of the efforts of all components of the national

economy. <u>Enhancement of the material and cultural level of
life of the population and further expansion of social production come solely from this net income</u>.

This is why the index of the profitability of the operation
of all units in the national economic system acquires decisive
significance. Consequently, one of the basic planning indices
must be the planned profitability target for the enterprises
and economic agencies. The index of profitability must be determined in percent of the fixed and circulating assets. In
planning profitability, a rigorous delimitation of two stages is
needed:

a) determination of the plan target with respect to profitability;

b) establishment of a scale of material reward for production personnel relative to the actual level of profitability (approximately the type of scale presented in Comrade Liberman's article).

The plan target for profitability should be defined in the
form of plan charges on fixed assets. These charges should be
designed to cover the expenditures of society upon expanded
reproduction of fixed assets, associated with the growth of the
social product.

The results of the economic activity of each enterprise
and combination of enterprises should be determined not only
by factory production costs, but also by total outlays of the
national economy.

<u>The time has come to eliminate the situation in which
fixed assets allocated by society to any given production entity
are given without charge</u>. Society constantly and continuously
reproduces its fixed assets on an expanded scale. And each
enterprise should participate in this process of reproduction
in proportion to the fixed assets granted to it by society to make
possible its functioning. This participation should be expressed
in standardized charges on fixed assets, just as the partial
participation of the enterprise in social expenditures is reflected in charges on wages.

The coefficients of reproduction of fixed assets with various material compositions are not identical. The material composition of fixed assets varies from one branch of industry to
another, and from one enterprise to another. Therefore, the
norms of charges on fixed assets should be differentiated by
branches and groups of enterprises, with consideration of the
material composition of fixed assets and the coefficients of

reproduction of various types of assets. These norms should be standardized over a ten to fifteen year period. If the plan norms of charges are changed from year to year, such planning of profitability can only have negative consequences.

It should become the chief concern of economists and planners to study the causes of failure of individual enterprises and branches to meet these long-term norms. Every failure to meet a profitability norm should be regarded as a loss and be reflected in the national bookkeeping as a grant or subsidy by the government to the given production unit. The major duty of economists and planners is to develop, in good time, measures that will eliminate such losses. In the majority of cases, improvement of planned prices and selection for the enterprise of an appropriate assortment of output would make it possible quickly to liquidate the operation at a loss of many branches and enterprises.

The most important task in reorganizing planning is to bring about a radical change in the situation that now usually comes into being in relations between enterprises and planning agencies. The enterprises must offer the planning agencies alternative plans for increasing their output capacity, and alternatives with respect to the volume and possible assortment of products, showing the cost to the economy as a whole, including planned charges on fixed assets in accordance with long-term norms. The planning agencies should select the enterprises to which it is economically advantageous to issue a particular order for commodity deliveries of various products.

It is necessary for each enterprise to seek to obtain from the planning agencies the largest possible plan target for commodity deliveries, in which case a situation opposite to that now existing will come into being. This is entirely possible if fulfillment of the order will be profitable not only to the state but also to the individual production unit.

Toward this end, the enterprise must possess a fund for material incentive, the size of which must depend upon the actual level of profitability. This attained level of profitability should include the actual payment of minimal planned charges on fixed assets, as well as economies in factory production costs calculated by comparison with the preceding period (on the condition that there is no reduction in quality of output).

This incentive fund should be calculated as a percentage of the difference between the level of profitability planned and actually achieved, the former being determined as the charges on the fixed assets in accordance with norms fixed for a long period. As in the scale presented by Comrade Liberman, these percentages paid into the incentive fund should be long-term norms, diminishing logarithmically as the level of actual profitability increases. Moreover, contributions of specific size to the incentive fund must be made from standardized receipts on account of charges on the fixed assets.

The incentive fund should be expended in accordance with definite rules: as bonuses to plant personnel and as an addition to the total wage fund, issued over and above basic individual pay rates. A portion of the incentive fund might be spent on improving the living conditions and communal life of the personnel of the given enterprise, particularly on housing construction and on bonuses in the form of apartments with greater conveniences.

The procedure for building and expending the entire incentive fund and that portion of it constituting the joint wage fund must be regulated by a special law of the Supreme Soviet of the USSR.

With such a system of planning material incentives for the personnel of enterprises, one may be confident that the relationship between planning agencies and state and cooperative enterprises will change radically. Under such circumstances it will not be the planning agencies that will press enterprises to accept particular plan assignments but, on the contrary, enterprises will seek to obtain the highest possible plan assignment from planning agencies.

It is usually held that under such a system enterprises will seek to obtain more profitable orders and avoid less profitable ones. But why should that which is profitable and necessary to the state be unprofitable to the enterprise? This occurs only because of defects in planned price formation and shortcomings in planned guidance. It is unfortunate that our planning agencies very often permit such powerful economic means for stimulating production as price planning and assignments from profits to the enterprise fund to escape their control. Quite often there is no need whatever to raise the price for products in short supply. It would do merely to increase the percentage of profit on such products that goes into the enterprise fund.

In funding means of production, under which material resources are dispersed among an enormous number of holders of innumerable physical funds, a "metabolic" sickness inevitably develops in the economic organism of society. The constant shortage of material resources is a manifestation of this disease of economic exchange. Such a situation is inevitable when, for example, marketable funds for surfacing materials and paints are doled out per million rubles of new construction and when it is necessary to begin to "time" these funds so that they do not come in response to that quarter million rubles expended at the initial stage of construction, i.e., when these materials are as yet virtually unneeded.

It has long been time to organize materials supply along the lines of state trade, and to cease to distribute commodity resources in accordance with the infinitely complex system of physical funds. Conversion to state trade in the field of materials supply would doubtless make it possible, in quick order, to eliminate the situation in which there is a constant shortage of material resources. This would also be facilitated by the planning of profitability, accompanied by elimination of the gratuitous provision of fixed assets and by introduction of long-term norms in the area of material incentive.

The need for a reorganization of planning has certainly matured. It is essential that the plan be fully harmonized with the principle of material incentive.

Management of the economy must be built on a proper combination of planning and material incentive — the two basic principles of socialist social development. Coordination of these social principles assumes the introduction of long-term plan norms, an increase of the enterprise funds (in particular, of the incentive fund), and expansion of the rights of enterprises in the expenditure of these funds.

Improvement of the price mechanism, introduction of long-term norms of profitability and of material incentive, expansion of the rights of the enterprises and abandonment of petty supervision over them — these are the principal landmarks to be followed in the reorganization of planning that has now become urgent.

Voprosy ekonomiki, 1962, No. 11

On Improving the Forms and Methods of Material Incentives*

B. Sukharevskii

A great deal of work has been done in recent years to improve the system of wage payment in the industrial branches of
the national economy. The adjustment of wages has strengthened
the principle of material incentives, made it possible to narrow
the gap between the wages of workers and employees in the
low- and high-wage brackets, and facilitated the introduction of
the six- and seven-hour working day without the reduction of
wages. All this has promoted the fulfillment and overfulfillment
of economic plans in the principal industries during the first
three years of the seven-year period.

Nevertheless the forms and methods of material incentives
cannot remain unchanged; they must be improved constantly.
The CPSU Program has posed new important tasks of improving the forms and systems of wage payment.

Of late many articles have appeared in the press which con-
tain concrete proposals for solving the new tasks raised by the
Party Program. Prominent among them is the article by E. G.
Liberman published in Pravda on September 9, 1962. It raises

---

*The article by Sukharevskii is the text of a report delivered
by the author at a session of the Learned Council on Economic
Accounting and Material Incentives of the USSR Academy of
Sciences held September 25-26, 1962. The papers of other
participants in the session follow.

The author is a member of the State Committee on Labor and
Wages of the USSR Council of Ministers. This is a discussion
article.

basic questions of material incentives in connection with problems of improving planning and the management of socialist enterprises. It is not surprising that the article has aroused considerable comment. Many of the proposals advanced in it have been supported, some have been challenged, others require further study and elaboration, or can be implemented only after the necessary prerequisites are created, especially, important changes in the field of price formation.

The National Economic Plan and the Socialist Enterprise. The improvement of economic management requires, first and foremost, the solution of the question of improving planning and of relations between the socialist enterprise and the national economy. Socialist ownership of the means of production makes it possible, with centralized management of the national economy, to ensure the unity of interests of the individual enterprise and the entire national economy, the individual worker and society as a whole, on the basis of the comprehensive mobilization of internal reserves in order to accelerate the rate of communist construction.

The problem is to make full use of these opportunities, to correctly balance centralized planning with the creative initiative of industrial executives and collectives, and to expand the rights of enterprises within the framework of the single national economic plan.

Shifting the center of gravity of industrial management to the economic councils has opened new opportunities for such initiative. The adjustment of wages, the introduction of a unified state wage rate system, and the greater role of technologically substantiated output quotas have created the prerequisites for raising the economic level of planning labor productivity and wages. But, as experience shows, these new opportunities are still being poorly realized and in some cases are even obstructing the utilization of the internal reserves of enterprises. (1) Why is it that in many cases the plan actually restrains the initiative of enterprises?

Plant workers criticize the economic councils because frequently the plans drawn up for the respective enterprises are not coordinated as far as production, construction, manpower, and material and technical supply quotas are concerned, the plan targets being distributed arbitrarily among the various enterprises; as a result the necessary stability of the plans is not ensured. This encourages the plant management to build up reserves against unexpected changes and to show extreme caution in forecasting possible rates of growth. The erroneous practice of planning "according to the level achieved" has also been rightly criticized. With concrete facts, enterprise personnel demonstrate how the initiative displayed by a plant which has undertaken to reach higher targets, then turns against that plant because the following year it receives still higher quotas, while those that have been lagging receive easier plans. Economic council officials, in turn, blame the planners and industrial executives for frequently failing to help the economic council find reserves and for trying to obtain from the state as much as possible for as little as possible. Both economic council and plant executives agree that the existing material incentive system should be changed so as to strengthen the unity of interests of the plants and the national economy in the best utilization of reserves.

It must be said that these mutual criticisms are justified. In order to solve the outstanding problems of economic development it is necessary to follow both roads and to coordinate measures for improving planning with those for improving the material incentive system. The best system of material incentives can be reduced to naught by incorrect planning. Material incentives for enterprise personnel can be fully ensured if a certain portion of the economic effect achieved by them is channelled for the benefit of the whole society, while another, previously agreed portion, is placed in the hands of the workers of the given plant, that is, those who contributed to the common cause of advancing the national economy. Planning according to the level achieved undermines these fundamentals of material incentives.

On the other hand, no matter how well planning is organized, it cannot take into account all the existing internal reserves. Therefore a system of material incentives which makes of the plan an "absolute," a single criterion for evaluating the plant's work, restricts initiative and makes the plan an obstacle to more rapid economic development.

What, then, is the way out of the situation? How can we carry out the instructions of the CPSU Program to make all enterprises interested in higher planning quotas?

The "Automatic Self-Regulator." A way out of the apparent contradictions has been suggested in our press in the form of a kind of automatic "self-regulator," which would encourage the enterprises to mobilize all their internal reserves and make rational use of their financial, material and manpower resources. The role of such an automatic self-regulator, it is claimed, can be performed by profitability (the ratio of profits to the production assets of the enterprise). The idea is that if a constant relationship is established between profitability and the material incentive funds built up by the enterprise, and if maximum premiums are provided for indices stipulated by the plan, the enterprises themselves will be interested in higher planned quotas. The introduction of such an index, in Liberman's view, will make it possible to expand the rights of enterprises and to reduce drastically the number of indices planned by the economic councils for the enterprises under their jurisdiction.

E. G. Liberman suggests that the economic councils should assign to the enterprises in their charge only the volume and assortment of commodities to be manufactured and the delivery dates; the enterprises themselves would then draw up their overall plans, which would include such indices as labor productivity, manpower, wage payments, reduction of production costs, accumulations, investment in new machinery. It will be readily observed that these suggestions are aimed at the fuller utilization of the law of value in the practice of socialist economic management. They proceed from the consideration that, if the stimulating role of economic levers based on the law of value is enhanced, the workers of an enterprise will themselves seek and find internal reserves. This will make it possible to relieve the enterprises of petty patronage and will ensure their economic rights and independence. In this respect Comrade Liberman is moving in the right direction.

However it does not follow from this that introduction of the profitability standard will make it possible to limit the plan indices for the enterprises to be stipulated by the economic councils to "quantity-assortment targets" alone. This part of Professor Liberman's suggestions arouses serious objections. The root of the mistake of this proposal lies in the fact that it ignores the unity of physical and value relations in social reproduction, and confuses the conditions of reproduction of a single enterprise with those for the national economy as a whole.

For as long as we are dealing with an isolated enterprise we assume that any labor saving achieved by it will later be embodied in a corresponding quantity of means of production and items of consumption required by the enterprise. If the enterprise increases its profit through economies and uses the money for investment, it is assumed that it will be able to obtain the necessary equipment and building materials. If an enterprise that achieves higher profitability raises the wages of its workers accordingly, it is assumed that they will be able to obtain a corresponding amount of consumer goods. But such assumptions hold good only as long as they refer to a single enterprise. As soon as we take the national economy as a whole, the possibility of providing every enterprise with the required means of production and consumer goods remains only an assumption. In real life guarantees must be created that every enterprise will receive the physical means of production and consumer goods. But the profit regulator does not provide such a guarantee even if the quantity-assortment targets are drawn up for every enterprise.

In the profitability controversy some economists have based their objections to making it a regulator of social production on the contention that profit is a capitalist category. Such objections, of course, are untenable, for in a socialist economy the essence of profit is entirely different. Actually, though, profit and profitability norms cannot act as automatic self-regulators primarily because they are purely value categories embracing only one aspect of the reproduction process.

Close links exist between enterprises and the national economy which, under conditions of commodity-money relations, appear in two forms: physical and value. For an economy to develop on a planned basis it is important to ensure proper proportions, both in value and physical terms. It is not accidental, therefore, that in addition to profitability, it is proposed to assign to enterprises the quantity-assortment targets. This, however, is not enough.

Would it be correct, for example, to refrain from assigning to enterprises their wage-fund quotas on the assumption that in their drive for greater profitability they will use the manpower and wage funds at their disposal with proper discretion? No, it would not. For greater profitability can be achieved in different ways: by economizing on material expenditures or by utilizing manpower. Heavy industry enterprises which increase the output of metal, fuel or machines create a basis for developing all branches of the economy, but they do not produce the consumer goods needed to provide the material counterpart of increased wages.

Consequently, not every increase of profitability will ensure the creation of the necessary material foundation for higher wages. In order to ensure the correct balance between money

income and available commodities, for any norm of profitability
a balance between money income and available commodities
must be established in a centralized manner; hence the need
for planning wage funds for enterprises from above.

Similarly, it is difficult to accept the proposal that enterprises invest at their own discretion, proceeding only from
their desire to increase their profit-to-production assets ratio.
Whereas production plans determine national economic proportions for the current year, investment plans determine the balance of production capacities in the economy and, therefore,
economic proportions for several years to come. How, then,
can an enterprise decide for itself what capacities should best
be developed for the long-term satisfaction of national economic
requirements?

Thus, centralized planning for enterprises cannot be limited
to the so-called quantity-assortment targets. At the same time,
though, enterprises must be relieved of petty patronage, and
other means must be found to encourage them to seek higher
planned quotas.

In the course of the discussion suggestions have been made
which point to a way of solving this task.

Three Criteria for Evaluating an Enterprise's Work. Whatever material incentive indices we take — volume of production, labor productivity, production costs, profit or profitability — there always arises the problem of the criteria according to which an enterprise's achievements are to be assessed.

In my view, there are three such criteria. The first is the
plan: fulfillment or overfulfillment of the plan for the corresponding index. The second is the criterion of dynamics: the expansion of output, the increase in labor productivity, the reduction in production costs, the increase in profit, in short, the degree of improvement shown by an enterprise in the given index.
The third criterion is the level with respect to the production
assets at the disposal of the enterprise and in comparison with
the achievements of other enterprises.

At present bonuses are paid to managerial and engineering-
technical personnel for fulfillment and overfulfillment of the
plan for reduction of production costs, and in some industries
for overfulfillment of production plans. Thus the measure of
the results of an enterprise's work is the plan, its fulfillment
or overfulfillment. (2)

In view of the shortcomings in planning, some comrades suggest that the material incentive system be divorced from the plan and that bonuses be paid regardless of fulfillment or degree of overfulfillment of the plan. Others propose proceeding from a comparison of the work of related enterprises, with the highest bonuses going to those which have achieved the best results in their group. Other economists consider the method of comparison to be impracticable and see a way out in a comprehensive study of the specific conditions of each individual enterprise and in assigning well-grounded plans.

In view of the difficulty of comparing the results of the work of related enterprises it is proposed that the evaluation of plant operations be based on a comparison of results not with other enterprises but "with themselves." The more rapidly an enterprise improves its work, the faster it forges ahead, these people say, the greater its material rewards should be.

Who is right in this dispute?

It seems to us that these proposals actually do not contradict one another. Their authors are simply viewing the same phenomenon from different aspects and the deficiency in their proposals is a one-sided approach to the question. Can material incentives be separated from the plan, if the plan determines the necessary proportions in the national economy? Obviously not. But this does not mean that bonuses should be paid for fulfillment of the plan regardless of the extent of improvement of plant operation that was stipulated by the plan and of the level achieved in its fulfillment. The point is to find a correct combination of the three criteria in assessing an enterprise's work.

In the material incentive scheme suggested by Comrade Liberman the basic criterion is the profitability achieved by an enterprise. Of course, if an enterprise raises its profitability, then according to his scheme it will receive a greater bonus fund. But since the formation of the bonus fund depends only on the profitability achieved, the enterprise will be able to receive substantial bonuses without improving its work as compared with the already achieved level. This is the drawback of the bonus scheme suggested by E. G. Liberman.

It seems obvious that it would be more correct to base the scheme not on the achieved level but on an index of improvement of plant operation, whether it be output, labor productivity, production costs, profit or something else. At the same time the bonus rate must be differentiated to take into account the achieved level for the given index. How can this be done?

Let us take a concrete example. Suppose in a given industry the bonus scale is constructed so as to depend on expansion of output. For a certain percentage of increase of output the enterprise would receive a certain sum for its bonus fund. This sum must be differentiated so as to take into account the level of utilization of production capacity. Say, in a given year output increased when the initial level of utilized production capacity was 70%; for each 5% increase a certain bonus rate would be established. But if the initial level of utilization of production capacity is not 70% but, say, 80%, then the bonus rate for the same 5% increase would be higher in the given year.

With such a bonus scheme the enterprise will be stimulated, regardless of the level already achieved, to strive for improved work each year, quarter and month. This is what is of decisive importance for the development of the socialist economy, the consolidation of its material and technical basis and the raising of the living standards of the people. At the same time material incentives will be linked with the level of production achieved.

Is it correct to make material incentives depend only on the improvement of work provided for by the plan (higher output, lower production costs, etc.) without taking into account the fact that one enterprise may be using its reserves poorly while another has achieved outstanding indices? Obviously, such a one-sided evaluation would be wrong. The decisive factor, of course, is our forward movement. But on the other hand it is easier to raise the level of utilization of installed capacity from 70 to 80% than from 98 to 100%. If only the indices of increased output and better work as compared with a corresponding period of the previous year (for output, or production costs or profitability) are taken into account, then the foremost enterprises which have already utilized their reserves "lying at the surface" will be placed at a disadvantage as compared with the lagging enterprises with untapped reserves. This, of course, would be wrong.

Payment according to work presumes a social evaluation of the measure of labor, which must rise constantly. The yardstick is the experience of the best enterprises, which must be widely disseminated. To evaluate an enterprise's work only according to the aspect of "comparison with itself" would mean to ignore the experience of the best. Tremendous reserves in the national economy exist because of the fact that there are large gaps between the performance of the best, average and lagging enterprises, and it is these gaps that must be bridged as quickly as possible. Therefore the method of comparing related enterprises is essential. This is the real way to implement the requirement of the Party Program which calls for greater material incentives for the enterprises to promote the most widespread application of advanced experience.

The question arises: what about incentives for those enterprises which failed to improve their work in a given year? For such enterprises there should probably be an approved scale of bonuses set up depending on the level of work achieved, which should be less than the bonuses due for a minimum improvement. Suppose an enterprise receives a bonus fund of 100 units for its workers for a 1% increase in output. If there is no increase, no improvement as compared with the previous period, the bonus fund should be less than 100 units. The size of the bonus should be in direct proportion to the achieved level of work: the higher the level, the closer the bonus fund approaches the 100 units.

Should the bonus be the same for fulfillment and overfulfillment of some index of the plan? Stimulation of the enterprise to undertake larger planned targets presumes differentiated bonuses for fulfillment and overfulfillment of the plan. Some suggest in this connection that the bonus for each per cent of increase over and above the plan be, say, one-half the bonus for each per cent of increase embodied in the plan. Comrade Liberman suggests that the bonus for the actually achieved index in the case of overfulfillment of the plan be fixed as an average of the actual index and the index provided for in the plan. In this case the enterprise which succeeded in being assigned a lower plan will get a smaller bonus fund than the enterprise in which the indices stipulated in the plan are closer to the actually achieved plan fulfillment. The question of the most efficient bonus system for overfulfillment of the plan must be studied further, but in any case the scale of material incentives for fulfillment and overfulfillment of the plan should vary.

The conclusion can thus be drawn that any bonus system should be connected with the plan, but the size of the bonus should differ depending on the growth (increase of work) stipulated by the plan and the achieved level of work as compared with other related enterprises.

Can such a problem be solved?

Long-Term Standards, Distribution According to Work, and Utilization of the Law of Value. In order to solve this problem standards must be devised for related enterprises and these standards should be long-term ones. No firmly based systems of planning or material incentives are possible without some standards.

What are the standards to be considered?

First of all, there are the standards which could be used to evaluate the level of work achieved by an enterprise. If the bonus is set so as to depend in some way on the utilization of production assets, then a standard must be devised based on the relation between output and fixed assets, or profit and production assets, a comparison of which makes it possible to evaluate the degree of utilization of these assets. Second, there are standards that could determine what portion of the additional income (economies from reduced production costs, profit, etc.) should be available for bonuses to the workers. Obviously, these standards must be of a long-term character, that is, for at least two or three years, so that the enterprise need not fear that by undertaking a higher plan in a given year it would place itself in unfavorable conditions in the following years. These standards should not be individual ones, but rather they should apply to groups of more or less similar enterprises. Advanced experience is still being slowly introduced in industry. Hence, the size of the bonus should depend on the introduction of advanced experience, on the level of work achieved in comparison with the foremost enterprises. This can be done if group, not individual, standards are introduced.

The necessity of drawing up group standards for related enterprises is also suggested by other important considerations. The point is that there is a certain contradiction between distribution according to work and distribution which takes into account the law of value. Application of group standards for related enterprises will make it possible to combine rationally distribution according to work with the utilization of commodity-money relations.

Actually, what we want now is to link to a greater or lesser extent material incentives with the general indices of an enterprise's work, such as its profitability and profit, which in turn determine the relation between prices and costs. As for price, it should approach closer and closer to the socially necessary expenditure of labor, that is, the expenditure of labor under average social conditions, average technological levels, average intensity, average skills, etc.

However, distribution according to work implies equal pay

for equal work in more or less similar conditions of production.
If wages were to depend directly on the degree to which the
means of production used by workers correspond to average
social requirements, this would be a violation of the principle
of equal pay for equal work, since enterprises differ markedly
among themselves in technical level (which basically does not
depend on the workers). As for payment according to the results of each individual worker, the question is solved comparatively simply. At enterprises with different technical levels
the same wage rates are usually in effect. Production quotas,
however, differ depending on the means of production used by
the worker; hence for different technical levels of production
there are different rates of pay for work done. Bonuses to
engineering-technical personnel, which are based on production
costs, are determined according to the reduction of production
costs as compared with the plan, with the cost plan being established separately for each enterprise.

The drawback of this system lies in the fact that the cost
plans drawn up for each enterprise separately are highly subjective in character. They frequently conceal or justify the lag
of some enterprise. At the same time an enterprise lacks incentives to improve its machinery since the results of such improvements will immediately be incorporated in the targets of
the plan for reducing production costs for the following year
or quarter.

Thus in practice the plan and material incentives are based
on individual conditions in individual enterprises, whereas the
output of these enterprises is sold at prices fixed on the basis
of average indices for the respective branches. Therein lies
the contradiction of the present material incentive system, its
weakness from the standpoint of utilization of the mechanism
of the law of value. Since it is based on individual indices and
not on group standards, it fails to develop the necessary incentives for lagging or average enterprises to strive to achieve the
level of the best. The progressive aspect of the law of value in
industry, it will be recalled, consists in the fact that it encourages the limitation of individual expenditures to the socially
required level and their reduction below the socially required
expenditures.

The solution lies in evaluating the work of enterprises and
paying them bonuses not on the basis of average or individual
conditions but by differentiating these indices according to
groups of enterprises with similar operating conditions. Then

the enterprises will be interested in improving their equipment and operating indices within their group, so as later to pass from a lower to a higher group where the bonus provisions are higher. (3)

In order to link material incentives with the economic indices of plant operation (production costs, price, utilization of assets) working time standards alone are insufficient. A system of differentiated technical-economic standards is required. In this sense E. G. Liberman's suggestions concerning the elaboration of technical-economic standards for groups of related enterprises are a step in the right direction.

In the future another step forward should be made: the prices according to which the enterprises market their output should be differentiated for groups of enterprises. Subsequently these prices would be levelled out on the scale of a whole industry and the consumer would receive commodities according to uniform prices. Such a solution of the problem would make it possible to connect, in an improved manner, distribution according to work with the economic indices of plant operation, with their profitability.

Some economists, seeking to link material incentives with the dynamics and level of plant operation, suggest the following solution: that the size of bonuses depend on the reduction of production costs of a given enterprise and on the degree of reduction of these costs compared with the average for an industry as a whole. If the reduction of costs at a given enterprise is greater than for the industry as a whole, then the bonus should be larger. If, on the contrary, a given enterprise has failed to keep abreast of the reduction of costs in the industry as a whole, then its bonus should be reduced.

However, with such a bonus system, first, the workers would never know in advance what bonus they might look forward to, since average indices for the industry as a whole are unknown to them and hardly depend upon them. Second, and most important, with such a bonus system the workers of an enterprise would be more interested in a lower average index for the industry, since the higher the average index for an industry as a whole the lower is the evaluation of the work of a given enterprise as compared with the industry as a whole and, hence, the lower the bonus. Thus, enterprises would be more interested not in spreading advanced experience but in retarding introduction. Obviously this proposal is unacceptable.

It goes without saying that the drawing up of standard indices will take time. But the question today is that of employing the

existing extensive network of research institutes and enterprises to organize the elaboration of technical-economic standards on a national scale and to make wider use of them for
improving planning and the material incentive system.

Incentive Indices. At present engineering-technical personnel and office employees receive bonuses for fulfillment
and overfulfillment of the plan for production costs and, in
some branches, notably in heavy industry, for overfulfillment
of production plans and improvement of quality. In some industries a part of the bonus is paid out depending on above-plan
profits (e.g., on state farms) or on achievement or surpassing
of planned incomes (e.g., in automotive transport).

The Party Program poses the task of stimulating higher
profitability of production, improving the utilization of production assets and installed capacity, and raising the yield of
each ruble of investment. What are the best ways of solving
this task?

E. G. Liberman suggests that a uniform bonus index be introduced in all industries and at all enterprises, which would
simultaneously determine the size of the bonus fund of the
respective enterprises. This index is the ratio of profits to
production assets. Like a number of other writers, E. Liberman regards profit as the main, decisive, and only measure
for evaluating plant operation. In the course of the discussion
other indices have been suggested for evaluating plant operation and material incentives: expansion of output, increase in
labor productivity, reduction of production costs, etc. It has
also been suggested that uniform indices of plant operation and
material incentives not be established for all industries, but
that we differentiate the indices by industries and even by
enterprises of the same industry.

It is therefore necessary to review, at least briefly, the pros
and cons of the various indices, their applicability to separate
industries, enterprises and their divisions, and also to different categories of workers.

In principle, the profit index, especially when regarded in
relation to the production assets of an enterprise, has a number of advantages over other indices, including production
costs. First of all, profits reflect changes in both the quantitative and qualitative indices of plant operation. Even if production costs per unit of output do not change, while output expands, profits increase. The reduction of costs, on the other
hand, does not depend as much on the volume of output as do
profits. Second, of course, profits depend on prices, and prices

provide, or should provide, a social evaluation of the expenditures of labor which society considers necessary at a given time or for a given group of enterprises. This is reflected in profits. The reduction-of-costs index, naturally, lacks this feature; in order to evaluate the level of production costs it would be necessary to establish a standard cost for various groups of enterprises. Third, the quality of output may also influence profits, if quality is taken into account in the price (grading, durability, etc.). Production costs as such are unable to reflect changes in quality. Profits relative to production assets are also an indication of the extent of utilization of the latter, which makes this index preferable to the index of output relative to fixed assets. The latter gives us some idea of the utilization of fixed assets, but it says nothing about savings of living labor. Profits, however, when related to production assets, also reflect savings of both living labor and materialized labor and the utilization of the assets themselves.

Thus in a socialist economy profits, in essence, may be a comprehensive index reflecting the savings of expenditure of social labor. That is why an increase in profitability is a requirement of socialist economic management in all branches of production. But profits are a purely value index reflecting savings of working time abstracted from the concrete forms of labor. But from the standpoint of the proportional development of the national economy, and at different stages, the expansion of output of one or another product in kind may vary in importance, and the effect of their increased output cannot be gauged only by the savings of working time at the producing enterprises.

While seeking to economize on labor in all industries, we should bear in mind the different role played by the various industries in the development of the national economy. The point is that higher output of the means of production will ultimately determine the possibility and extent of labor saving in all branches of the economy. Therefore the effectiveness of greater output of the means of production is determined not only by the saving of labor in the industries producing them but also by the saving of labor in the industries using them. It goes without saying that everything must be done to reduce production costs in heavy industry. But of primary importance for the country's economic development is the increase in production itself, the growth of output in heavy industry, which is the decisive factor in the saving of labor in the national economy as a whole. On the other hand, in some industries expansion of output at present cannot be accepted as the primary index for material incentives for they are limited by raw material supplies. In these industries the expansion of production can be achieved mainly through further economizing of raw materials. In some of them, savings of working time are even less important than savings of raw and other materials.

Thus, the application of the index of profitability in different industries is of different significance, and this calls for a differentiated approach in choosing the material incentive indices for the respective industries. In this respect the considerations voiced in Pravda by correspondents of Ekonomicheskaia gazeta are completely justified.

Neither should the practical difficulties of applying the profitability index under conditions of the existing price system be overlooked. This consideration is one of the reasons why the present system of bonuses for engineering-technical personnel and office employees is based on the index of production costs, not profits. Under the present price system the price not only takes into account the conditions of production but also serves to stimulate a certain structure of consumption. Therefore profits reflect both the results of production and the requirements of consumption. For the profitability index to become the basis of material incentives the consumption factor must be excluded from the price at which the enterprise markets its goods. Furthermore, at present prices are established according to the production costs plus profits scheme, with profitability being calculated as the ratio of profits to production costs. If material incentives are to be established on the basis of a certain relationship of profits to assets, this should be taken into account in the price structure as well.

In addition, some industries and enterprises now operate at a loss, and the profitability of some products fluctuates sharply. This necessitates an adjustment of the sales prices of manufactured goods. Existing prices, finally, are based on average costs in an industry and, as a rule, prices adapted to separate groups of related enterprises are not used.

In view of this it would seem expedient to improve the material incentive system in two stages. In the first stage, measures should be taken that can be carried out before a general adjustment of prices is undertaken. In the second stage, measures should be taken together with the price adjustment. Of course the index of production costs is also affected by prices. However the effects of prices on costs are less than on profits. Therefore in the first stage it would be correct to make wider

use of the reduction-of-costs index. Furthermore, it is necessary to introduce partial changes in price formation, especially
for new commodities, and to assess fulfillment of the plan for
the actual assortment of commodities not only according to
costs but to profits as well. This will make it possible to make
wider use of profits as a basic or supplementary index for
gauging material incentives.

Thus it seems possible to distinguish four groups of industries according to the basic index of material incentives.

In the first group, comprising primarily some of the extractive branches of heavy industry, the basic index will be
volume of output; in the second group it will be labor productivity; in the third, production costs; in the fourth, profitability.
As mentioned above, in determining the measure of reward
each of these indices should be evaluated from the standpoint
of improvement as compared with a preceding period, the
achieved level (in comparison with production assets and the
level of indices shown by related enterprises), and also fulfillment or overfulfillment of the plan. It may be necessary to pay
bonuses for two indices (as is the current practice in some
industries), including, for example, the quality index.

However ideal the incentive scales for a group of enterprises
may be, they cannot take into account all the concrete operating
conditions or the sum total of the indices of plant operation.
Economic councils should therefore be allowed to increase or
cut bonuses within certain limits, taking into account the
sources by means of which the indices were achieved, in particular, whether expansion of output was achieved through increasing labor productivity or employment, as a result of investment from the state budget or the introduction of new capacity from the enterprise's own accumulations, etc. All this
would serve to extend the rights of the economic councils in the
choice of material incentive indices for the various enterprises
under their jurisdiction.

Material incentive indices for executives, engineering-technical personnel and office employees should be linked to material incentive indices for workers. Thus if engineering-
technical personnel in some industry are to be paid bonuses for
fulfillment of production plans, taking into account the increase
in production stipulated by the plan and the level of capacity
utilization, then the bonus scheme for workers will have to be
changed, for in some industries the latter are currently paid
bonuses for fulfillment and overfulfillment of section plans,
regardless of the production expansion stipulated in the plan

or the extent of utilization of production capacity in that section.
Workers' bonuses can also be differentiated depending on the
increase and level of labor productivity.

The Bonus Fund and the Combination of Individual and Collective Incentives. In improving the forms and methods of material incentive it is important to ensure the correct combination of individual and collective incentives. The possibility of
such a combination is inherent in the very nature of wage payment. One of its basic features is that, as a portion of the national consumption fund, it should, as Lenin pointed out, also
depend on the overall results of the operation of a given enterprise. This is possible because in the manufacturing industries
wages are paid not out of the budget but out of the proceeds of
sales of commodities produced by the given enterprise.

The combination of individual and collective incentives requires a differentiated approach to material incentive indices
for different categories of workers within an enterprise, taking
into account the nature of their work and its effect on the overall results of plant operation. All categories of workers should
be rewarded to some degree or other for the improvement of
the overall results of plant operation, and the size of such
bonuses should be increased. The basis for bonuses paid to
every group of workers should be the indices over which they
themselves have the greatest influence.

The profit index, which in some conditions may serve as a
basis for determining the bonus fund of the enterprise as a
whole, cannot be brought down to the shop or section. Even the
production cost index, as actual practice shows, cannot in all
cases be brought down to the shop or section. Experience shows,
however, that other indices can be adopted for the lower divisions (volume of output, economies of specific materials, higher
labor productivity), which in the final analysis determine savings from reduced costs and the profitability of the enterprise
as a whole.

If in evaluating material incentive indices for the different
plants the rights of the economic councils should be broadened,
it would seem logical to broaden the rights of plant executives
in choosing the incentive indices for various categories of
workers within the given enterprise.

Many economists consider that a basic solution of the problem of material incentives lies in the creation of a single bonus
fund at enterprises for all categories of workers, for all in-

dices of work and for technological improvements. Such suggestions fail to take account of two considerations. First, the overall results of plant operation (such as profits and savings due
to reduced production costs) depend to a varying degree on the
efforts of workers of different categories.

It would be wrong to make the bonus paid to a worker for
better work depend only on the profits of the enterprise as a
whole. This would not ensure the correct combination of individual and collective incentives. Workers should be paid bonuses
for the results of their work from the wage fund and, in addition, depending on the work of the enterprise as a whole. The
building up of a bonus fund for managerial and engineering-
technical personnel can and should be made dependent on the
overall results of plant operation (profits, savings from reduced
costs, expansion of production with due regard for the utilization of assets).

Second, in solving the question of material incentives for
the creation and introduction of new machinery, it should be
borne in mind that enterprises develop machines "for themselves" and machines "for others." If a machine-building plant
masters and expands the manufacture of machines for other
enterprises and industries, then obviously the measure of material incentives should be dependent, as it is today, on the
economic effect derived from the use of these means of production in the respective industries.

Thus the material incentive fund cannot be uniform; enterprises should have specialized funds (a bonus fund for engineering-technical personnel and office employees, a bonus fund
for new machinery).

On the one hand, the bonuses paid to engineering-technical
personnel within the limits of a special fund will create a definite guarantee of payment of the allocated bonuses to the
workers involved and, on the other, it will ensure a correspondence between the overall sum of bonuses in the national
economy and commodity resources.

In the course of the discussion it was correctly noted that
experiments are necessary not only in technology but in economics as well. The carrying out of an experiment in the sphere
of planning and in the material incentive system bears directly
on the vital interests of the workers and employees of the enterprises where the experiment is to be launched. Its positive
or negative results may influence the fulfillment of their plans
by the respective enterprises. Therefore any experiment must

be preceded by thorough calculations, with due consideration given to the suggestions and advice of workers of the enterprises where the experiment is to be launched. From the elaboration of the basic direction in the improvement of planning and the system of material incentives it is necessary to move on to practice. This will make it possible to mobilize new reserves to accelerate communist construction.

***

From all that has been said the following conclusions may be drawn concerning the further improvement of the material incentive system for enterprise personnel.

1. Measures to improve the material incentive system should be carried out in two stages. The first, under the existing price system; the second, together with the improvement of prices. In the first stage the profitability index should be used to a greater extent in socialist economic management. In adjusting the price system it is necessary to take full account of the requirements stemming from the utilization of the profitability index for material incentives for workers of enterprises. Improved planning, especially of labor and wages, is a prerequisite for effective measures aimed at improving the forms and methods of material incentives.

2. Enterprises can be made to seek higher plan targets primarily by a correct combination of three criteria for evaluating plant operation and material incentives in paying bonuses to managerial and engineering-technical personnel; these are the plan, the dynamics and the level of work of the enterprise. The bonus scales should be based on the improvement of plant operation (expansion, dynamics), with a differentiation of bonuses depending on the level achieved and on whether this level was stipulated by the plan or exceeded it.

3. It is necessary to work out technical-economic standards of a dual nature: (a) standards according to which the level achieved by a given enterprise in some index can be judged relative to the enterprise's production assets and to the level of work achieved by other enterprises, primarily leading enterprises; (b) standards determining the measure of incentives depending on the indices of plant operation. These standards, as a rule, should be not individual or average, but for groups of enterprises working in similar conditions and with similar technological structures (not technological levels but technological structure; thus, modern steel works and old plants in the Urals cannot be grouped together).

The standards should be long-term ones, covering two or

three years, so as to guarantee that the enterprises will be interested in mobilizing internal reserves and in receiving higher planned quotas.

4. It is necessary to ensure a unity of principles of material incentives while simultaneously taking into account the specific features of the economics of an industry and its role in advancing the national economy. The basic indices of material incentives for managerial, engineering-technical personnel, and office employees should be differentiated according to groups of industries. The basic index should apparently be: (a) for industries not limited by raw material supplies (e.g., the extractive industries) and possessing an extensive market: the expansion and level of production; (b) for industries with a largely similar output: the growth and level of profitability; (c) for other industries: the reduction and level of production costs. In all cases the improvement and level of utilization of production assets should be considered (total assets, fixed assets, production capacity). Wherever profitability cannot serve as a basic material incentive index some additional incentive should be introduced for profitability.

Bonuses for fulfillment and overfulfillment of planned quotas should be differentiated, being less for overfulfillment and more for fulfillment of the plan indices.

5. Material incentive indices for various categories of managerial, engineering-technical personnel and office employees should also be differentiated. The rights of the economic councils should be extended in the evaluation of material incentive indices for enterprises in their charge, and the rights of plant directors in choosing the bonus indices for various categories of enterprise personnel within the framework of the general material incentive scheme.

6. Engineering-technical personnel and office employees should be paid bonuses from a special fund, while workers should be paid bonuses for individual and collective results by a team or shop from the wage fund. It does not seem expedient for an enterprise to establish a single bonus fund to include bonuses for new machinery.

7. In order to encourage enterprises manufacturing new products, we must expand the number of industries in which bonuses are increased, depending on the share of new output (as is the case at present in the machine-building industry), as well as the industries in which the new technology development funds are used to cover expenditures on the manufacture of

new equipment; in revising prices we must ensure greater profitability of production of new commodities.

Footnotes

1) This was well demonstrated in an article by the chief accountant for the Abagursk Sintering Plant of the Kuznets Metallurgical Combine, Comrade Kargopolov: "Barrier or Springboard" [Bar'er ili tramplin], published in Izvestia. Of course, the plan should not be a barrier, but a springboard for progress.

2) For enterprises whose plans do not call for a reduction in production costs (for reasons not of their own making), lower scales of bonuses have been established.

3) In instances when the average norms can be attained without large-scale capital investments, uniform average branch norms are applicable.

Voprosy ekonomiki, 1962, No. 11

The Main Question: Criteria for Premiums and Indices Planned for Enterprises

I. Kasitskii

Our meeting is discussing fundamental problems of our work, difficult and controversial problems. Let us consider three of them: 1) the criteria and conditions of material incentives, and whether universal solutions are possible; 2) the planning of indices for enterprises and the character of the indices proposed by Comrade Liberman; 3) the connection between material incentives and liability.

During the war, when enterprises were given the right to establish their own premium systems, a machine-building plant at which I worked established two indices for most shops: the commodity volume index and the labor requirements index. Premiums were awarded for surpassing the previous level provided the plan was fulfilled, that is, for the results actually achieved. Several shops were given other targets, depending on their nature. Thus, shops with a complete production cycle were given production cost targets. With the introduction of this premium system the situation with respect to planning changed radically. The shop superintendents began to search diligently for reserves, not a single rationalization proposal was shelved if it could yield the slightest advantage, more attention was given to cutting losses, etc. The plan was fulfilled regularly. The factory was regarded as one of the best in the region. This was not only the result of the premium system; of course, many other steps were taken to improve equipment, operating techniques and production organization. This example, or economic experiment, as it is now called, made it

---

The author is associated with the Committee on Economics and Production of the USSR Council of Scientific and Technical Societies.

135

possible to formulate and raise the question of improving the
premium system not on the basis of academic speculation but
on the basis of concrete experience.

This problem has a long history. As far back as March
1946, the question of extending the rights of directors and improving economic accounting and the system of material incentives was discussed at a meeting of economic executives
at the Dzerzhinskii Engineers' Club. Later a number of articles
on this subject appeared in the magazines Kommunist and Voprosy ekonomiki and the newspapers Pravda, Izvestia and Ekonomicheskaia gazeta. All this shows that the problem has long
been an urgent one and has undergone extensive discussion. It
is obviously no accident that the Academy of Sciences has set
up the Learned Council on Economic Accounting and Material
Incentives.

In recent years important changes have taken place in the
premium system for engineering-technical personnel, a system
of material incentives for new machinery has been introduced,
etc. The time has come, however, to proceed with a radical
solution of this serious, urgent and very important problem on
the basis of the instructions of the 22nd Party Congress and the
new Party Program.

Let us consider the question of criteria and conditions of
material incentives. Is the awarding of premiums for fulfillment and overfulfillment of the plan justified? It is my profound conviction, which is shared by other comrades, that it is
here that the cause of some of our major shortcomings lies. On
this question we are fully in accord with E. Liberman, although
we do not agree with him on many points. The plan cannot serve
as the criterion for awarding premiums.

Let us recall some of the points in the Party Program. It
says, in part, that it is necessary to raise the role and incentives of enterprises in the introduction of new techniques and
the fullest utilization of productive capacity. The Program also
points out that economic management should rely on material
and moral incentives to achieve high production indices. Furthermore the Program says that the whole system of planning
and evaluating the work of central and local organizations,
enterprises and collective farms should make them interested
in higher planned quotas. And, finally, the Program points to
the need to strengthen collective forms of material incentives
and to heighten the concern of each worker for a higher level
of work of the enterprise as a whole. Nowhere in the Program

or in a single speech at the Congress is anything said of the
need to award premiums for fulfillment or overfulfillment of
the plan.

Can we say that our plans embody all these points? I think
not. Of course, some comrades think that plans are never artificially reduced. In a review of letters and articles Ekonomicheskaia gazeta (June 12, 1961) cited the head of the technical
department of the Khabarovsk Economic Council, G. Grigor'ev,
who declared that there are no cases of violation of the principle of personal material interest or of economic executives
concealing reserves or making exaggerated claims. Such
charges, according to Comrade Grigor'ev, cast a slur on plant
executives. The newspaper remarked, quite rightly, that the Grigor'ev statement meant that either he did not know the facts of
life or that he wished to varnish them. Hardly anyone today
really thinks that if an enterprise receives a premium for
overfulfilling the plan it will seek to have that plan raised.

In Ekonomicheskaia gazeta some comrades referred to the
need to introduce a standard plan, that is, quotas on the basis
of scientifically substantiated norms. But this can give rise to
attempts to keep norms down, and you may be sure that plenty
of scientific reasons will be found for this. Other comrades
think that premiums should be awarded for fulfillment of the
plan, taking into account its degree of difficulty. But is it possible to determine the difficulty of an enterprise's plan? In an
editorial review by Ekonomicheskaia gazeta published in Pravda
on September 24, 1962, it was stated that this was quite
"simple": all that had to be done was to transform coefficients
into percentages, multiply the volume of output by the increase
in labor productivity and profitability, and divide all this by the
coefficient of production cost reduction. Mathematically and
economically, this is a sheer artificial indicator which does
not take into account capacity or new output, that is, the main
elements that make for a difficult plan. Since these factors are
quite specific for every enterprise and are very difficult to
define, no such index can be established.

As for the idea of not taking the plan as a criterion for
premium awards, in the 1962, No. 3 issue of the magazine
Sotsialisticheskii trud (in a review and a paper by Comrade
Shkurko), it was claimed that this did not correspond to the
principles of socialist planning. We think, however, that fulfillment of the plan must be a prerequisite for premium awards,
for ours is a planned socialist economy. This prerequisite,

however, leaves room for other premium indices.

Our differences with Comrade Liberman begin with the question of what to award premiums for, although his article in Pravda marks a serious step forward. Now, for instance, he agrees that enterprises should be given planned targets for specific commodities and amounts of output, something he did not propose in his article in Kommunist in 1956.

Any attempt to make a system universal would be a profound mistake. On this score the Ekonomicheskaia gazeta review in Pravda mentioned here is quite right. For how can one establish identical indices for a metallurgical and a machine-building plant and for an enterprise of an extractive industry, coal, or oil, for example? How can one establish identical indices in the machine-building industry, say, for an enterprise with well-organized, mass, and more or less stable output, and for an enterprise with individual-type output with frequent changes of items, for a new and an old enterprise, etc.? This would be absolutely incorrect of course. That is why it is wrong to attempt to introduce uniform indices, whether they be profitability, as Comrade Liberman suggests, or something else. They must be carefully differentiated not only by industry but also depending on the type of enterprise, on the nature of its work, its possibilities, the utilization of its reserves and its technological level. The right to establish them must be given to the economic councils. If the premium system suggested by Comrade Liberman were introduced at all enterprises, or even at groups of enterprises, we would be committing a grave mistake.

It is appropriate to examine some of the indices which Comrade Liberman suggests for awarding premiums, and in the first place an index such as the production program. Liberman does not mention the basis for establishing this criterion. Nevertheless there are objective criteria here. In some cases this may be production capacity, in others, the output of new commodities, etc. Liberman's article fails to answer this question.

Finally, there is Comrade Liberman's main index: long-term profitability norms. It will be recalled that at the 22nd Party Congress Comrade Khrushchev spoke of the need to raise the role of profit and profitability. This is undoubtedly true, but what is meant by profitability? Profitability has always been defined in economic practice and literature as the percentage relationship of profits from sales to production cost of the goods sold. Comrade Liberman views this index as the ratio of net profit to the sum of current and fixed assets. What would occur if we introduced this index, which is widely used in capitalist countries (for this is neither more nor less than the rate of profit on invested capital)? First of all, the enterprises would be interested in a reduction of the denominator. As far as current assets are concerned, this would make sense (although Comrade Liberman for some reason ignores circulation funds). But with respect to fixed assets, this would lead to technological stagnation: in their drive for a higher index the enterprises would stop improving their machinery and would reduce investment in fixed assets. The interests of the national economy would clash with the temporary and specific interests of the workers of an enterprise.

E. Liberman says that only the production program should be planned. But ours is a planned economy; what about such an index as the balance of labor resources? If we lost control of manpower planning, the acute problem of the utilization of considerable manpower reserves would soon arise. Incidentally, it is E. Liberman himself who speaks of freeing manpower. How is this manpower to be employed? This question is solved on the basis of a manpower balance. This cannot be ignored. Or can we fail to plan the wage fund? How could we then regulate currency circulation, strengthen the ruble, regulate retail trade? These indices cannot be ignored.

Let us consider net profit, which Comrade Liberman links with the premium system. As we know, this is made up of income from sales less production costs, plus or minus so-called non-marketing proceeds and losses. At many enterprises these constitute a fairly large sum and frequently they do not depend on the enterprise, while affecting net profit. Hence they will affect the amount of premiums. Is it correct to proceed from net profit? No, it is not.

It should also be remembered that income from sales is a function of prices. And although E. Liberman mentions shortcomings in the system of price formation and notes that the process should be more flexible, we must take into account the fact that this is not yet the case. The revision of wholesale prices of means of production currently being carried out is not a final solution of the problem. Therefore some enterprises may have very good indices, and hence high premiums, regardless of how they worked.

Well, then, for what should premiums be awarded? It

seems to us that an answer is given in the Party Program, which notes that in our country the firm law of economic development is in force: the achievement of maximum results in the interests of the people with minimum expenditure. In some cases this may be profitability (this index cannot be rejected wholesale) or the ratio of profits to production costs; in other cases it may be the index of maximum utilization of production capacity, but most frequently it should be the index of production costs. In our view this would be the most correct approach. But in any case premiums should be paid within the limits of the plan and for the actual achievement of minimum costs and maximum results. In this there is no contradiction between the interests of a given enterprise and the interests of the national economy. Premiums must be paid for fulfillment of the plan only in relation either to a previously attained level or to some other standards established for several years ahead. Although long-term standards do not seem to have worked well in Czechoslovakia and Poland, we should examine how they were introduced: maybe there were some faults. Any idea can be ridiculed. But it should be obvious that some degree of stability of standards and indices creates much greater incentives than if they are changed a dozen times a year.

Characteristically, almost all the articles, except for a few published in Kommunist, speak only of incentives and say nothing about liability. But Lenin spoke not only of material incentives but of material liability as well. Incentives and liability are two organically related elements of economic accounting. They cannot be divorced; they must be tackled together.

Finally we come to the organizational forms of solving the question under discussion. We should not be hasty. It must be solved gradually, in stages: first the question of criteria for premiums and indices to be planned for enterprises must be settled; then a system of economic experiments in different branches and at different types of enterprises must be carefully devised. It will take a year or even two to carry out all these experiments and the necessary theoretical research. Only then, after the results of this work have been summarized, can concrete proposals be made. These are problems of immense political and economic importance, and they must be solved fundamentally, on the basis of thorough study.

Voprosy ekonomiki, 1962, No. 11

Against Oversimplification in Solving Complex Problems

A. Zverev

The Party and Government are carrying out a comprehensive program of democratization of state and social life. Important changes in the management of industry and the reorganization of management in agriculture have increased the role of the working people in economic management and in all other spheres of our country's state and social life. This is one of the important factors determining the steady rise in the effectiveness of social labor, the successful progress along the road of communist construction.

Further improvements in the utilization of labor, material and financial resources in production, a consistent reduction in expenditures of social labor per unit of output and a steady increase in labor productivity constitute a law of the socialist economy, the importance of which has grown still more in the period of the creation of the material and technical basis of a communist society.

Economic incentives for increasing and improving output, personal material incentives for the working people to improve the results of production and economic activity (in combination with moral factors) play a highly important role in implementing production and other economic plans as far as both quantity and quality are concerned. At present the state spends several billion rubles a year on premiums to workers at enterprises and other economic organizations. In addition, about 80% of profits over and above the plan remain at the enterprises. Con-

---

The author, former USSR Minister of Finance, is associated with the Institute of Economics of the USSR Academy of Sciences.

siderable sums allocated to enterprise funds from planned profits are utilized for improving material standards and cultural services for the workers of these enterprises. Thus tremendous sums are spent on providing material incentives for workers to improve and increase output. If these sums are correctly utilized, much can be achieved to heighten the interest of workers in the effectiveness of production.

Unfortunately, there are still many shortcomings and violations in the utilization of these funds. There are also shortcomings in the system of building up premium funds and in the indices which are used to evaluate the results of the work of an enterprise, a group of workers and individual workers, as well as in the indices according to which premiums and other material rewards are paid. It is necessary to improve in every possible way the system of material incentives for improved production indices and results of economic activity. Material and moral factors should be used fully in solving the problems of expanding socialist production and in raising its efficiency.

It goes without saying that economic planning must also be improved in all its aspects, freeing it from many minor faults and eliminating excessive patronage over economic branches and, all the more so, over individual enterprises. Excessive, uncalled-for interference of planning and other bodies in the work of economic branches and enterprises should be stopped. The state plan should be freed of unnecessary indices and provisions for petty patronage. This will benefit planning and economic management. It will make it possible to strengthen the fundamental role of state planning in the development of the socialist economy in the period of the full-scale building of communism and will improve the quality of economic planning.

If the problems raised by E. Liberman in his article in Pravda are approached from these positions, we may note certain positive aspects. The author attempts to identify the correct criteria for the evaluation of plant operations, criteria which would facilitate the best utilization of productive forces and raise production efficiency. In any case, the questions posed by him have attracted widespread attention and sparked a lively discussion of many vital problems. Undoubtedly this will considerably accelerate the search for, and elaboration of, the best methods of material incentives.

As for the basis of the concept which E. Liberman advances, it seems to me to be dubious, insufficiently thought out and inconsistent. He takes an oversimplified and sketchy approach to the solution of an extremely complex problem.

Any innovation should be implemented only if it is really progressive, if it moves things ahead and facilitates the development and improvement of production and raises its efficiency. What are Comrade Liberman's new proposals? First, he suggests that to "improve" planning we should no longer plan from above enterprise targets for reduction of production costs, increases in labor productivity, utilization of raw materials, fuel and other supplies, that we stop planning employment figures for enterprises and wage funds. All this should be planned by the enterprises themselves. The only centralized targets handed down to the enterprise should be the volume and composition of production. Second, he believes that the enterprises should decide for themselves the amount and direction of investment. Third, he thinks that we should stop drawing up yearly plans of profitability and profits, and establish average, long-term profitability standards according to an appropriate centralized scale. Fourth, E. Liberman suggests that enterprises and their workers be awarded premiums for fulfillment of the fixed profitability standard, with higher premiums for overfulfillment of profitability indices. In his view the best utilization of all manpower, material and financial resources of an enterprise, the rational distribution of capital investments and their greater effectiveness will be ensured by discontinuing centralized planning of investments and production cost quotas, and by introducing a stable profitability scale which will be fixed for an enterprise for a long period as a percentage of total fixed and current assets.

It should be noted, first of all, that E. Liberman falls into a contradiction. He bases his new system of planning on the premise that the State Planning Committee and the economic councils are less informed as to the capabilities of the enterprise than the enterprises themselves. At the same time he considers that the state plan and the plans of the economic councils should specify volume and composition of production, as well as the suppliers and consumers of the finished product. The question then is, how can the State Planning Committee and the economic councils establish the volume and composition of production for enterprises if they know the production capacities so poorly? Actually this is not the case. The planning organs and the economic councils are obligated to know, and actually do know, the production capacities of enterprises. Without this they would be unable to plan and the economic councils would be unable to direct the work of enterprises.

E. Liberman claims that his proposed system of planning and material incentives precludes the possibility of understatement and "concealment" of plant capacities from the planning authorities and the economic councils. However his system of material incentives does not preclude the possibility of enterprises understating their production capacities so as to overfulfill the production plans and the fixed profitability rate, thus receiving more premium funds.

Furthermore, E. Liberman suggests that the enterprises themselves draw up their investment plans, except for "large-scale" investments. But investments are the main aspect of the process of enlarged socialist reproduction. Furthermore, it is necessary to ensure the necessary proportions in developing different branches. Under socialism this balanced development must be reflected and grounded in the state economic development plans. It can best be achieved when the amount and direction of investments are determined by a state plan and not decided by an enterprise. The enterprises are ignorant of the various national economic interrelations, and even if they wanted to establish the balance of the economy they could not do it. Even now we have mistakes in planning that result in disproportions in the national economy; what would happen if every individual enterprise would engage in this work? There can be no doubt that it would give rise to still greater mistakes, with more serious discrepancies in planning investments and more serious disproportions in industrial development.

In the USSR the development of the productive forces, and of industry in particular, takes place not within territorial or national boundaries but according to the principle of the best utilization of natural and other resources so as to achieve the greatest effect from investments for the benefit of the society as a whole and each member individually. An example is Kazakhstan, where the Soviet state has been investing material and financial resources far in excess of the national income produced by the republic. This is of advantage to the state and the people. Implementation of this extremely important principle of Soviet economic policy, which ensures a steady rise in the efficiency of social production, is possible only if state planning discipline is observed, only as a result of better planning on a national scale, not by transferring the planning of investments from the State Planning Committee to the economic councils and enterprises.

We should not forget that some republic, regional and plant

executives display parochial tendencies. The results are misappropriations of investments, disruption of economic balance and disorganization and weakening of the planned basis of the economy. The ending of capital investment planning by the State Planning Committee and the economic councils, and the transfer of these functions to the enterprises, would only intensify these negative tendencies and in no way improve the utilization of resources for capital investments and the observance of balance in the development of the economy.

It is necessary to improve state planning, including the planning of investments, to accelerate the commissioning of new enterprises and to raise the economic effectiveness of investments. At the same time the rights of the republics, economic councils and enterprises in economic management must be observed.

E. Liberman's views on production costs seem very strange. If we are to believe him, it is not planned cost assignments and control on the part of the respective organizations and authorities over their fulfillment that will guarantee the fulfillment of quantitative and qualitative indices but the average rate of profitability. How can we end the planning of cost targets for enterprises? If we discontinue the centralized planning of production costs, what use will the economic councils have for these indices? And why then should the State Planning Committee consider problems of costs if the indices drawn up by it will be obligatory to no one? We must not forget that production cost is the basic qualitative index of any production plan. The effort to reduce production costs means a daily search for ways and means of reducing the expenditure of social labor per unit of output.

The same must be said about the planning of labor. The correct distribution of manpower resources in the economy is a prerequisite for the growth of the national income. The increase in labor productivity determines the possibilities and rates of enlarged socialist reproduction, the rise of the material and cultural standards of the working people and the implementation of other state and social tasks. High labor productivity and low production costs are the basis for the development of the productive forces of the socialist society at a higher rate than under capitalism. Comrade N. S. Khrushchev noted that the higher the labor productivity and the lower the production costs, the faster will be the rise of living standards, the greater the accumulations of the economy, and the faster we forge ahead to our cherished goal, communism. The steady growth of labor productivity on the basis of technological improvement, the better organization of operating techniques and the systematic rise in the skills of workers are prerequisites for the building of communism. High rates of enlarged socialist reproduction and a steady improvement in living standards are inseparable from a more rapid growth of labor productivity than of wages.

Tremendous importance must be attached to the correct and economical utilization in production of raw materials, fuel, power and other supplies. All these elements are linked with production costs and expenditures. The economical utilization of all these elements raises the effectiveness of production, reduces the expenditure of living and materialized labor, increases accumulations in the economy and, consequently, the reserves for further enlarged reproduction and satisfaction of the growing requirements of the population.

Comrade Liberman suggests that all these great and difficult problems be solved by a simple method: by introducing a long-term average rate of profitability for every enterprise, a material incentives scale, and by discontinuing the planning of investments and cost assignments for enterprises.

Obviously, to improve the organization of production and promote technical innovation, to steadily increase labor productivity and reduce production costs, it is necessary to improve the system of material incentives for better work. But for this purpose the power of state planning must be utilized. Taken by itself, the average rate of profitability established for enterprises for a long period, and the material incentives scale drawn up by the author, cannot solve the great and difficult problems of production. These questions require a thorough study and a premium system which would really raise the effectiveness of social production.

It seems to me that in his proposals E. Liberman oversimplifies the problems and views them too schematically. In his view the premium scale should be drawn up in a centralized manner. But such a scale can be drawn up only on the basis of production costs. Consequently the average rate of profitability for the scale must also depend on costs. The only difference is that at present production costs are studied for each enterprise and cost reduction and profitability targets are fixed with due consideration for all circumstances on the basis of production conditions. The author, however, suggests that the

average profitability for a group of enterprises be found, that
is, a quantity which will not correspond to the objectively possible profitability of every enterprise in the various groups of
the scale. As a result, some enterprises will be receiving
higher profitability quotas, which will serve to reduce their
premium funds, while other enterprises will be overfulfilling
their targets because of understated profitability rates and will
receive unduly high premiums. This shows how erroneous is
the scale of average profitability indices and premiums.

It should not be forgotten that in industry one enterprise or
another is constantly improving its machinery, organization and
techniques. This drastically changes labor conditions and raises
productivity. Conditions of fuel and power supplies, the quality
of raw materials, haulage distances of raw materials and supplies and many other conditions change. In many cases these
changes may not depend on the enterprises; yet as a result
their actual profitability may change. What is to be done? Comrade Liberman might say that in such cases it would be necessary to revise the profitability rates. That is so, but in view of
the tremendous scale of our industry and the constant changes
taking place in it, the revision of profitability indices of enterprises will be a mass and continuous activity. It will be carried
out by some special authorities with whom the enterprises and
other organizations will inevitably engage in unnecessary disputes. In revising profitability it will be essential to take into
account the production costs of each enterprise under conditions
in which these will be determined by each enterprise for itself.
Moreover, some organization will have to establish at what
enterprises the profitability indices must be changed in view of
new conditions. Those enterprises in which the conditions of
achieving the established profitability have worsened will immediately raise the question of their revision, while those in
which the profitability is higher may not "realize" that it is
time to ask for a revision to raise their profitability quotas.

Is such an "innovation" necessary or useful? I think not.
It is impossible to replace the role and power of state planning
by establishing an average profitability rate for all enterprises.
This is an "innovation" which can be detrimental to the economy.

Discontinuing the planning of production costs for enterprises would reduce the attention paid to this basic economic
question of production; in any case, control over the production
expenditures of enterprises would be reduced. Average profitability cannot replace the control functions of higher organs.

Disputes would be inevitable in establishing new profitability rates. The possibility of receiving higher profitability quotas and the related possibility of industrial executives concealing their reserves are not precluded, particularly since the profitability quotas would be set for long periods.

Elimination of planned production cost quotas for enterprises would make it more difficult, and in some cases impossible, to plan the net income of society and finances, for the enterprise which sets its own costs would be unable to take into account possible changes in production conditions and would proceed from the profitability indices established for a long period. Difficulties would appear in drawing up the overall financial plan of economic development; additional difficulties would appear in the distribution and redistribution of the national income on a branch and territorial basis; the role of the State Planning Committee, the economic councils and financial organs in planning costs and bringing the quotas down to the enterprises would be reduced, which would serve to weaken to some extent the drive for higher production efficiency.

In conclusion, it should be noted that the author's understanding of profitability and profit contradicts generally accepted theoretical concepts, according to which profit is the main part of the surplus product created by the workers' surplus labor. According to E. Liberman's conception, it seems that profit is created not only by the workers' labor but also by the fixed and current assets. It is hardly necessary to prove the erroneousness of such a "theory." In questions of price formation, which constitute a large and independent economic area, the author's ideas lead to the conclusion that the methodological basis of price formation in a planned socialist economy should be the price of production, which is characteristic of the capitalist system of economy.

These are by no means all the considerations against the advisability of accepting E. Liberman's suggestions. However, the facts above are sufficient to reject his theory and his recommendations.

Voprosy ekonomiki, 1962, No. 11

Making Enterprises Interested in More Intensive Plans

V. S. Nemchinov

The chief objective of our discussion is to reveal the common ground that unites the proponents of different views on the question under discussion. If we are to engage in endlessly piling up our differences we will not accomplish anything. Judging by the speeches made at the conference, there is agreement on the thesis that while preserving the principle of centralized planning it is necessary to create opportunities for the display of local initiative, to take account of the different conditions in which particular enterprises operate. None of us holds that this basic thesis should be revised. Although A. Zverev said at the conference that E. Liberman had departed from the principle of centralized planning, this is not correct. Comrade Zverev's remarks indicate that E. Liberman's point of view has been misunderstood. Nobody proposes to revise or, all the more, to question the decisive role of planned targets, or the very principle of planning. It is absolutely unquestioned and obvious to all that planning is the basic principle of our economic system. But the question arises: how can this principle be most effectively applied? It should not be thought that a plan, no matter how ideally it has been drawn up, balanced and coordinated, should be handed down to individual enterprises in an immutable form. If we embark on this path we shall bind the enterprises hand and foot.

It is absolutely correct that with respect to calculation indices needed for drawing up a plan, still greater detail is necessary, including the drafting of balance sheets of an inter-branch

149

nature and in physical terms. However it would be erroneous to think that the entire system of such indices could be brought down to the depths of the economy and made obligatory for individual enterprises. Such practices would deprive the enterprises of all initiative. The working out of a plan is a matter of calculation; its implementation is quite another thing. A plan in the making is coordinated with other, related plans; it embodies definite directives. Its carrying out, however, also requires taking into account the conditions in which the enterprise operates. A plan presented to an enterprise should contain a minimum number of indispensable planned indices which would only regulate its activities, while at the same time allowing for the influence of all available economic levers. Moreover we must be able to coordinate all the elements of the system of centralized planning. As the experience of the Laboratory of Economic and Mathematical Methods of Elaborating a Model of the Planned Economy of the Belorussian SSR has shown, this is no easy matter.

We must not reconcile ourselves to preserving a situation in which a plan is drawn up in an atmosphere of competition of a sort, with enterprises trying to prove that their potentialities are small, and with planning organs seeking to set the enterprises as high targets as possible, higher norms, etc. These relations should be quite different: it is not the planning organs that should be seeking to set the enterprise as high target figures as possible, but the enterprise itself must strive to obtain from the state as intensive a plan as possible.

This can be achieved, in our opinion, if both the interests of the state and of the enterprises concerned are met. This could not have been done in the period of industrialization, during the years of war or of the postwar rehabilitation of the national economy, but it is quite feasible now. What is needed is a situation in which the planning organs become organs distributing orders, while enterprises seek to obtain these orders.

Another important question is the mastering of new output. The state allocates large sums for building new enterprises, and the newly built enterprises should be placed in conditions that will interest them in the maximum possible utilization of their equipment, in the full employment of their manpower and reserves. This can be done, in particular, when profitability is planned.

The shortcoming in planning just mentioned is not a defect of the system, but rather of the planning practices connected with planning intermediate, rather than final results.

The planning of profitability is based on a correct understanding of production costs as national economic expenditures. A. Zverev said at the conference that in this way we were reviving the capitalist category of the price of production. I am not in favor of using this category, but we must take account of value. It operates in our country, although the form in which it does is another matter.

Of course, a single average rate of profit is unacceptable in our conditions. On the other hand we impose special charges on wages, charges that make up the social insurance fund. As for the suggestion to introduce special charges for fixed assets, some comrades see in this something like establishing "interest on capital." On the same grounds we could consider the population's income from savings-bank deposits as profit on capital. Certain factories were mentioned at the conference, in particular, the Magnitogorsk Plant, which make considerable profits. However if one analyzes the extent to which work is mechanized at this enterprise, what its fixed assets are, the picture will not be so bright.

We cannot agree with the proposition that fixed assets should be cost-free. In striving for the expansion of production (for reproduction on an enlarged scale is a law of the socialist economy), why should we not take account of the resources that are used for this expansion? We have stopped at the categories of simple reproduction, believing it sufficient to include depreciation in production costs. What about society's expenditures on enlarged reproduction? How should they be distributed among enterprises? Who is to bear their burden? Only in the past could this problem be solved with the help of the turnover tax. Today the situation is different and we must not stop at that stage. As we know, there are coefficients of enlarged reproduction of fixed assets, coefficients which differ according to the latter's material composition. These coefficients can be calculated. The enterprise should be obligated to render an accounting of the full national economic cost, including special charges on fixed assets, rather than of a partial cost.

The above theses, in our view, disclose the point of profitability planning. The latter cannot be planned as a percentage of production cost price; it should be determined with due account taken of the extent to which the production process is provided with fixed assets.

Further, a premium fund should be set up at enterprises

by deducting a certain percentage of profitability, including deductions from savings on factory production costs, as the latter were understood previously. E. I. Kapustin has correctly remarked that quite a different situation will arise if this incentive fund is placed at the disposal of the enterprises. We paid attention only to indices for individual bonuses in solving the question of the sources of the incentive fund. However, even with the best possible set of indices taken as the basis for premiums, the indices may not be fulfilled unless the material incentive funds are properly established. It is not advisable to depend fully on the economic council in this respect, as suggested by Ekonomicheskaia gazeta. What is needed is to give the enterprises the right to choose their system of indices (and there should be several systems of indices) within the funds available to them, provided the plan is fulfilled and progress is made. The indices of progress are, undoubtedly, the decisive indices. But if the system of indices is made dependent on the fulfillment of the plan, then (since the plan will continue to change), as before, this will do it harm rather than good. Unfortunately, some comrades, A. Zverev and E. Kapustin in particular, do not agree that premiums should not be made directly dependent on the fulfillment of the plan. We proponents of this point of view do not claim that we are the only ones who know the truth. We must find a correct solution to this problem, search for it boldly, without looking back, without fearing absurd accusations on the part of comrades who are still living in the past.

It is alleged that E. Liberman and other comrades, who are advocating that the incentive fund be based on profitability, are in favor of a universal index. This is not correct. We are not for a universal index, but we believe that the enterprise requires a single source for its incentive fund. We should establish what unites us with E. Liberman and not what disunites us. This common ground, in our opinion, is the recognition of the need for setting up a single incentive fund for the personnel of the enterprise. Many people will probably also agree with the suggestion that the present fund of the enterprise be divided into an incentive fund and a reconstruction fund, and with the opinion that the enterprise should have its own means for the payment of premiums, and also with the proposition that it should be able to choose any of the incentive systems established by law. The practice of petty tutelage should be eliminated.

A system should be worked out which would make the enterprise itself demand a more intensive plan, one linked with the conditions requisite for its fulfillment. It is no secret that with a certain assortment of output the enterprise is prepared to tackle a double or triple plan. The planning organs should decide to whom to assign this or that task, and who will cope with it more successfully. The immutable law of economic activity formulated in the Party Program should manifest itself in full measure in this field.

Voprosy ekonomiki, 1962, No. 11

Important Condition for Improvement of Planning

G. Kosiachenko

The 22nd Congress of the CPSU has posed important tasks in the sphere of improving the management of the national economy. Hence the need to improve planning and material incentives for industrial enterprises. These two problems are closely connected, for any principles that are employed as a foundation for material incentives, as well as any bonus systems, will be useless unless they are connected with the formulation of the enterprise's plan and with the quality of this plan.

The main shortcoming in the field of planning is the neglected state of norm-setting. Norm-setting, which was unsatisfactory even earlier, has deteriorated in recent years. Hence the weak technical-economic foundation of the plans established for enterprises, the lack of coordination between production and supply plans, the constant alteration of plans — all of which leads to a lack of confidence in them among enterprise workers and undermines their attitude to the plan as law, as a directive that should be carried out unconditionally.

In order to eliminate these shortcomings it is necessary, in our opinion, to improve norm-setting. In his speech at the 22nd Congress of the CPSU, Comrade N. S. Khrushchev said that it is necessary to have "progressive planning norms for the utilization of all types of instruments of labor, raw and other materials, for technological methods and time limits for various jobs; it should be law for every manager to introduce these norms and abide by them strictly."

It is impossible to prepare a well-grounded plan without proper norms. It is said that economic councils prepare un-

The author is Director of the Financial Research Institute.

satisfactory plans for enterprises, compiling them by mechanically working out targets based on the level of production
attained by the enterprise in the preceding year. However, this
is explained not only by the lack of sufficient qualifications of
the economic council staff but also by the fact that they do not
have at their disposal the necessary instrument that would enable them to prepare a more firmly grounded plan. Progressive norms are such an instrument. They should be calculated
not for each enterprise separately but for a whole branch of
industry, or at least for a group of similar enterprises whose
technical level of production is more or less the same. Such
norms should reflect the achievements of leading enterprises
and promote the advance of lagging enterprises, orienting them
in such a way that they will make fuller use of their capacity,
more economical use of raw and other materials per unit of
production, etc.

The question arises: who should work out the norms? As
we know, this job is being done by the Research Institute of
Planning and Norms. However a single research establishment,
even if it does have the support of several branch research
institutes, is unable to solve this problem. This work should be
directed by Union and republic organs, agencies that are more
closely connected with the actual work of management and
planning of industry. The branch departments of the State Committees of the Council of Ministers of the USSR, in particular,
could render great aid in working out the technical norms for
the corresponding branches.

In order to eliminate the shortcomings in economic planning it is also necessary to introduce proper order in the
material and technical supply system; this should also be based
on technically substantiated norms, balance accounts, and an
improved supply organization.

In working out plans we must ensure a concrete approach
to each enterprise and take into account its specific features
in the coming planning period. This means that in each individual case it is necessary to provide for measures, which should
be taken by a given enterprise if, for example, it is undertaking
the production of a new item, or is entering the stage of mass
production of this item. It is also necessary to consider the
degree to which technical norms have been adopted, the extent
to which the enterprise is provided with raw materials for the
coming year, how its production capacity is being utilized, etc.

But such a concrete and comprehensive assessment of

all the conditions of an enterprise's work is possible only if a number of plan indices are worked out. That is why there can be no single universal index which could be used as a basis for assessing the work of an enterprise. Even such a summary index as the level or growth of profit relative to the preceding period cannot replace all the other indices in assessing the work of an enterprise, for profit, as we know, also includes elements of redistribution of net income.

Some economists propose that enterprises be granted bonuses not for the fulfillment and overfulfillment of the plan, but for the actual growth of output or the actual reduction of production costs in comparison with the preceding year. These indices are certainly of basic importance in assessing an enterprise's work, but if used alone they do not give us a correct picture of the enterprise's work, for they do not reflect the concrete conditions under which the results in question were achieved.

Proposals are also being made to grant enterprises bonuses, depending on the way in which a given enterprise fulfills or overfulfills the norms of utilization of production assets and labor set for the branch. The degree of fulfillment of technical norms is, naturally, an important index in assessing the work of an enterprise. However there are either no such norms or very few of them. That is why the job of working out norms should now be tackled in real earnest. But this is not all. In applying uniform branch norms at a given enterprise, to assess its work we must also introduce substantial corrections, taking into account the enterprise's specific features: its technical level, composition of output, etc.

Some economists are of the opinion that in order to make general norms applicable to the peculiarities of individual enterprises we must differentiate branch norms and work out group norms. There is no doubt that group norms are necessary. However average norms should not be replaced by individual ones. In evaluating the work of an enterprise, a concrete approach is required. For this purpose we must use a definite system of indices, rather than a single one, however important it may be. It is necessary to proceed from such a system in preparing a plan and in assessing the results of its fulfillment. Furthermore, in addition to certain common indices for all branches (for example, the fulfillment of the plan for composition of output and production costs), it is necessary to establish specific indices for each branch of the economy. All

these indices can be embraced only by a plan. That is why only
the plan proper and the degree of its fulfillment are generalized
indices on the basis of which an enterprise's work can be assessed. Consequently bonuses should also be granted for the
fulfillment and overfulfillment of the plan. We must raise the
significance of the plan. It must be a mobilizing factor but also
realistic. It is only on the basis of a plan that a bonus system
can be set up, a system that will play a substantial role in
achieving economies of living and past labor.

At the same time we must improve individual plan indices,
particularly value indices, and apply them with more attention
to specific features, taking into account the special characteristics of economic branches and endeavoring to have them reflect as precisely as possible the actual achievements of each
enterprise.

We must also find a more effective method of utilizing
bonus funds. In particular, it is our opinion that it would be
wise to grant larger bonuses for the fulfillment of plans than
for their overfulfillment; correspondingly the allocations to the
enterprise fund should be larger from planned profits than
from above-plan profits. It is important to raise somewhat the
role of individual bonuses from the enterprise fund. It is also
desirable to increase incentives for certain categories of
workers for raising the quality of output, for mastering the production of new types of goods, for improvements in production
technology that ensure substantial economies.

The plan should also reflect more fully the degree of utilization of fixed assets, particularly the time limits for mastering production capacity at new enterprises. We must establish higher depreciation rates at enterprises with superfluous
fixed assets and treat these as expenditures connected with the
enterprise's work, i.e., as losses. This will reduce somewhat
the zeal of managers prone to hoarding.

We must raise the role and responsibility of the heads of
enterprises and economic councils for the fulfillment of profit
plans. Among other factors this should be one of the criteria
for granting bonuses to certain categories of managerial personnel. In general it is necessary to define more precisely the
indices whose fulfillment would give the right to receive a
bonus. It is clear that these indices cannot be absolutely identical for all branches of industry.

The whole system and conditions of material incentives
should reflect in the fullest possible way the specific features
of production in different branches of the economy. For this

purpose it is necessary to raise the role of the economic councils in assessing the work of individual enterprises and in granting bonuses with due regard for the degree of difficulty of plans.

In analyzing E. Liberman's proposals we have come to the conclusion that he actually denies the significance which a plan has for an enterprise and makes everything depend on changes in profits. It follows from his conception that it is hopeless to raise the quality of the plan for an enterprise and therefore it is not worth attempting to improve plans. The only means possible, according to him, is to have the enterprise share in the profits and to set for it only the volume and composition of production. All other indices are to be worked out by the enterprises themselves in such a way as to ensure maximum profit. This is the only criterion which, in Comrade Liberman's opinion, will ensure automatically and by itself the full utilization of fixed assets and the economical utilization of raw materials, fuel and other materials per unit of output, and the wage fund, and will promote a rational allocation of capital investment, etc.

N. Antonov, one of the supporters of this conception, develops it in his article in Pravda (issue of September 14, 1962). He declares that profit should be the basic and the most important index, and not a derivative of the volume of output production costs. This means that production will be subordinated to changes in profits. In this case a wide range is opened up for the law of value and, rather than bringing any benefits, this will only harm the planning system.

In this connection we must also examine the proposal made by Academician V. S. Nemchinov, who considers that the time has come to replace the funding of materials by trade. True, it is wise to shift to trade in certain goods or materials, whose supply meets the requirements of the national economy. But given the relatively limited quantities or even lack of substantial reserves of many major materials, we cannot change over to their distribution by trade rather than by direct authorization.

Academician V. S. Nemchinov feels that the main reason for the shortage of materials lies in shortcomings in the supply system. True, shortcomings in the material and technical supply system make shortages more acute, but they are not its main cause. It is sufficient to cite, for example, the resources of agricultural raw materials which still restrict the growth of certain branches of the light and food industry, the resources

of metal which restrict the scope of production in certain
branches of machine building, etc. A shift to trade in metal
will inevitably lead to a somewhat different trend in the utilization of metal compared to that envisaged in our annual and
long-range plans. This should not be forgotten.

Let us take another example from the sphere of construction. In planning the material and technical supply of construction projects, preference is given to projects of special importance to the national economy. However there are also non-
centralized investments from local funds. The sum total of these
funds is greater than the volume of capital construction which
could be supplied with material resources. There are two possibilities: either material funds (not counting local resources)
should be allotted at reduced norms for non-centralized construction, or the volume of non-centralized construction should
be restricted and brought into conformity with existing resources. As a result, in the first case there would be a sharp
increase in the volume of construction and, as a consequence
of this, a scattering of resources and a growth in the number
of uncompleted construction projects. This is unacceptable.
There remains, consequently, the other way — that of ensuring
strict conformity between the volume of non-centralized investments and material resources. The question arises: how will
such conformity be ensured if, as Comrade Liberman suggests,
investments connected with the reconstruction of existing enterprises (not counting new ones) are to be determined by the
enterprises themselves? It can be said with complete certainty
that this will lead to a marked deterioration of capital construction in the country. However capital construction is a
national economic problem and not an isolated issue concerning
only individual enterprises. That is why it must be solved in a
planned manner, in close connection with national economic
targets and proportions established for the economy as a whole.

We must also examine the question of planning the wage
fund. E. Liberman holds that no targets for labor productivity
and increased wage funds should be set for an enterprise, for
it will naturally be interested in lowering production costs and
raising profits. However an increase in profit can be achieved
by economizing on past labor, while permitting a considerable
growth of wages. As far as the reduction of production costs
and increases in profits are concerned, it does not make any
difference whether the latter increased as a result of econ-

omies of past or living labor, but for the national economy it does make a difference. This question hinges on proportions whose significance goes far beyond a single enterprise: it depends on the relationship between Department I and Department II of social production, between accumulation and consumption, between money incomes and expenditures of the population, etc.

In this connection there is another question of great importance. Suppose that the additional growth of wages (compared with the calculations used as the foundation of the national economic plan) is accompanied by a corresponding additional increase in production. But in this case it is important to know at what enterprises this takes place — at those which produce consumer goods or at those producing means of production. In the first case the growth of money incomes of the working people will be accompanied by a growth in the market commodity stocks of goods; in the second case (particularly in branches producing the means of production for means of production) the growth of the wage fund in excess of the planned increase in turnover of goods and services can lead, in the absence of sufficient reserves of commodity stocks, to certain pressures and unjustified currency issue. Consequently, additional market stocks of goods will have to be found for the additional wage fund. That is why under no circumstances should we reject the regulation of the growth of the wage fund on a national scale and permit enterprises to solve this problem themselves.

Voprosy ekonomiki, 1962, No. 11

E. G. Liberman: Right and Wrong

K. Plotnikov

A great many concrete proposals have been made in the course of our discussion, and among these were recommendations diametrically opposite to Comrade Liberman's views. Apparently it is necessary to select those of the recommendations which conform in the greatest degree to the present stage of development of our national economy and provide a scientific grounding for them. All the proposals can be subdivided approximately into three categories: the improvement of planning of enterprises' production operations, the determination of criteria for the assessment of their work, and the improvement of the bonus system for workers. In his speech at the 22nd CPSU Congress, Comrade N. S. Khrushchev said: "Life itself calls for scientific grounding and economic calculations of a new, far higher order in the fields of planning and management." That is why the proposals made should, first of all, be assessed from the standpoint of scientific grounding and economic calculations.

Regarded from this standpoint, some of the recommendations do not hold water, first of all, because the problem under examination is not always considered in all its diversity and in organic relationship with the group of problems determining the relations between enterprises and the national economy. Without such an approach it is impossible to solve the problem of improvement of planning and management of the economy.

Even a capable manager, if he does not have the rights necessary for tackling the problems confronting him, will fail to solve them together with the workers of the enterprise. That is why the improvement of the enterprise's planning work is

---

The author is Director of the Institute of Economics of the USSR Academy of Sciences.

connected with the provision of greater rights for its director and with making the workers materially interested in the
results of their work. Unfortunately, many recommendations
in the proposals of some writers, including Comrade Liberman, are unsound in our opinion. They seek to solve the problem of planning in isolation from the national economy as a
whole and by disregarding the connection between the plan's
quantitative and qualitative indices. However we know that an
organic unity of these indices is one of the basic requirements
of planning. Their separation automatically leads to a contradiction between the demands made on an enterprise and its
possibilities. This can be illustrated by a series of examples.

Think, for instance, what would happen if the enterprise
itself determined the capital construction index instead of proceeding from the general, national economic tasks set by higher
organs. In order to answer this question it is sufficient to recall that with the existing system of planning and management
of construction organizations the volume of incomplete construction for the whole of the USSR amounts to 76%. Moreover,
instead of decreasing, this figure is tending to grow. Some
economic councils, proceeding from purely local interests,
curtailed allocations for the construction of particularly important projects and used the resources released thereby for
projects of secondary importance. Unfortunately they continue
this erroneous practice to the present day.

Perhaps all this was caused by excessive centralization of
planning and construction work management? Nothing of the
kind. We have about 10,000 building organizations, 50% of
which are subordinated to economic councils and local soviets.
Consequently the management of construction work is carried
out mainly by the local soviet government organs. Now just
think what would happen if 200,000 enterprises would begin
solving, on their own, the question of what and how they should
build. In that case the state would simply be unable to conduct
national economic planning and to observe the necessary proportions in the development of social production.

The absence of necessary centralization in capital construction planning and management of construction work already entails a great loss to the national economy. When staff
members of the Institute of Economics of the USSR Academy
of Sciences inspected 39 machine-building plants, it emerged
that at many of these the capital investment per unit of output
in reconstruction work was much higher than in new construc-

tion. For example, at the Voskov Tool Plant in Sestroretsk the capital investment per unit of output in new construction amounted to 81 rubles per ton, while similar investment in reconstruction work amounted to 106 rubles per ton. Approximately the same situation was observed at some of the enterprises in Cheliabinsk, Magnitogorsk, etc. The question of the relationship of capital investment per unit of output in construction and reconstruction of enterprises is one of the problems of major economic significance. It calls for a uniform solution conforming to overall state interests.

Therefore, in discussing the extension of the enterprises' rights, we should consider the major problems of capital construction. In this connection it should be recalled that 18 months ago the Economic Research Institute of the State Economic Council of the USSR and the Institute of Economics of the USSR Academy of Sciences prepared a draft for the improvement of capital investment planning indices. The draft was discussed at an all-Union conference; an understanding was also reached concerning the basic indices that should be introduced, but the question, however, remained unsolved. I think that the Learned Council for Economic Calculation and Production Incentives should become interested in this draft, which contains a number of important economically substantiated proposals.

The question of the basic criteria for assessing the work of enterprises is of great significance. Comrade Liberman is of the opinion that profit is the main criterion for assessing the work of an enterprise. Is this actually so? It seems that this is not entirely the case. Profit is indeed a very important index used for assessing the work of an enterprise and rewarding its workers. However this is an individual index, and a value index at that. It should not be forgotten that such a synthetic index is subject to the influence of many factors acting in different ways.

Let us see what the utilization of profit as the basic and only criterion of assessing the work of an enterprise would actually lead to.

The Economic Research Institute of the State Economic Council inspected 100 machine-building enterprises and observed marked variations in the level of profits of individual enterprises (the variations ranged from 5 to 60%). Can we conclude from this that the enterprises whose profits reached 60% worked well, while those with profits of only 5% worked poorly?

No, we cannot. Even one and the same item sometimes yields different profits for different enterprises. An enterprise's profitability is also influenced to a substantial degree by prices. Thus the high prices of goods produced by enterprises do not stimulate the utilization of agricultural machinery in collective farms. For example, the local economic council planned for the Sibselmash Plant in Novosibirsk a profit of 132% per cent of the production of tractor disk harrows. This is a direct result of the high sales prices of the goods. Such a high profitability, determined by the price list, reduces the desire of an enterprise's workers to strive for the utilization of all their capacities and, above all, for the reduction of production costs.

We cannot deny the importance of such an index as the profitability of an enterprise. However it should not be regarded as the main and, especially, the sole factor in assessing the work of an enterprise. An enterprise's work should be assessed with the help of a set of indices, both in value and physical terms. Only in this case will an all-sided assessment be made which will correctly reflect an enterprise's work. In particular, as an index in physical terms it would be suitable to use marketable output in both comparable and current prices, dispensing with the gross output index. This, however, does not completely solve the problem. It is important that the indices be differentiated according to branches of industry and types of production.

Stability is a very important feature of planning. It is impermissible to have production plans changed several times a year. The production plans prepared by enterprises should stimulate their workers to adopt higher targets, and here the system of material incentives is called upon to play an important role. It is our opinion that in examining this question one should not approach it from the standpoint of rejecting all the existing forms of incentives, as is being done by Comrade Liberman and some others. The important thing is not to reject as useless all the existing bonus systems but to eliminate from them everything that interferes with stimulating high indices of production.

The shortcomings of the existing bonus system lie, first of all, in the fact that bonuses are awarded regardless of the degree of difficulty of the plan. The bonus funds are not made dependent on the results achieved relative to the preceding period; a greater percentage of profit goes as bonuses for the overfulfillment of the plan than for its fulfillment. It is very difficult to receive money for setting up the enterprise fund. For this purpose it is necessary to fulfill the targets for the reduction of production costs, gross and marketable outputs, the composition of output, labor productivity, deliveries to other areas, state deliveries, the introduction of new machinery, the quality of output, etc. The failure to fulfill even one of these conditions makes it impossible to have allocations made to the enterprise fund. Thus its stimulating role is reduced.

In our opinion we must improve the methods of accumulating and utilizing the enterprise fund instead of rejecting it completely. Our discussion shows that most people are inclined to the opinion that the main foundations of the existing bonus systems do not have serious shortcomings. The point at issue is mainly the elimination of the obstacles interfering with the utilization of these systems for stimulating social production.

Problems of improving planning cannot be examined separately from problems of improving the organization of economic management and, unfortunately, not enough attention is devoted to the latter. It should be noted that in some large capitalist countries there are schools for training business executives. For instance, Harvard University (USA) has a school for business executives, training personnel for 300 concerns. There is no special training for managerial personnel in the USSR, although a great deal of attention was devoted to this matter in the 1930's. It is high time to tackle this problem seriously and to do so on a scientific basis. It is necessary to furnish a scientific groundwork for a system of organization of management of an enterprise's work, and to train systematically cadres of managerial personnel, setting up for this purpose the necessary facilities and the appropriate colleges and schools.

Voprosy ekonomiki, 1962, No. 11

Incentives Must Be Linked with the Long-Term Planning of an Enterprise

L. Al'ter

The broad discussion that is now under way on the issues of planning and material incentives reflects, as it were, the need to improve the system of planning our national economy, a need which has naturally arisen lately. Important tasks have been posed in the process of building the material and technical basis of communism. Among them, in the actual practice of planning, there is the problem of thoroughly substantiating plans from the standpoint of the economic effectiveness of capital investment and new equipment. There is an urgent need to combine planning, economic efficiency and economic stimulation of production into an integral process. It is also essential that planning be brought into full conformity with the new forms of management in the national economy. The main problem today is to apply improved scientific methods of planning, including mathematical methods.

We are faced with the following question: along what lines should the planning of the national economy by improved?

First of all, in improving scientific planning in conformity with new historical circumstances and new tasks, it is essential to overcome two unhealthy tendencies. One of them treats planning as an act of administration, as a volitional command. Here the problem of effectiveness, of the economic substantiation of the plan, and the issues of economic incentives are underestimated. True, we have already overcome this conception of planning to a considerable degree, but there are still some remaining elements of it to be noted, and they must be

---

The author is associated with the Economic Research Institute of the USSR State Economic Council.

eliminated at all cost both in theory and in practice.

The attempt to make broader use of economic levers and economic stimuli in planning is a healthy reaction against the administrative conception of a plan. It must be regarded as basically correct. However the attempt to impart a more profound economic content to planning often involves conceptions that lead in another direction. Instead of making use of economic accounting, profit, price and money in the interests of scientific planning, some economists are trying to invent an automatic system of categories that will function by itself and will perform a certain kind of "self-regulation" of the economic process.

In our opinion this is wrong. To a certain extent this is related to the conceptions that are called "institutional planning" in the West. This actually involves adjusting certain monetary and credit categories, following which the whole system will, as its inventors see it, work by itself. Incidentally, in France, where this system is called "indicative planning," it does not exclude the need to exert influence on investment as well.

These two unhealthy tendencies — first, divorcing planning from the actual economics of production and economic laws, an administrative conception of planning, and, second, transforming economic levers into a kind of automatically functioning mechanism — are incompatible with the system of state-controlled, planned management of the national economy, which is based on the principles of democratic centralism. The problem must be put as follows: it is highly essential to substantiate planning more profoundly by means of the categories of economic effectiveness and economic incentives, but not through an automatically functioning mechanism.

In discussing the proposal to revise planning on the basis of long-term standards of profitability, we must first answer the question: what does this really imply? Does this mean improved planning of the national economy or the substitution of unified national economic planning by something else, perhaps institutional regulation? The suggestions that are made in this connection, for instance E. Liberman's, contain a number of positive features, particularly the emphasis on the role of profit. We often underestimate the importance and role of profit, sometimes not coordinating it with the general system of plan indices. Sometimes people even say that profit is a capitalist category and that there is no need for it. These cannot be viewed as serious considerations. In the same way we might proclaim money and price to be capitalist categories as well.

Another suggestion that should be supported and utilized is that of introducing long-term standards for the material incentive system. Usually, when incentives are being discussed, they are linked with current planning. However the idea of long-term standards for economic stimuli provides the opportunity of connecting incentives with long-term planning.

A number of suggestions are based on the sound idea of the need to stimulate high, so-called intense plan assignments. The scale of long-term operation should not only stimulate high plan assignments by having the bonus fund allocated to the enterprise according to a fixed percentage relative to the plan, but should stimulate its interest in overfulfilling the plan as well. This is a very useful and rational proposal that should be applied.

In addition to these there were some unacceptable proposals made in the course of the discussion. This applies, for instance, to the suggestion that the enterprise should not be given any plan assignments with respect to labor productivity, production costs, wage funds, capital investment, supplies, introduction of new equipment, etc.

The role of profits can and must be raised; long-term standards should be introduced into the material incentive system, and high plan assignments must be stimulated. But these three elements should not be viewed as an automatically functioning mechanism, as factors which eliminate the need for planning the most important indices of an enterprise's work that we have spoken of above.

Indeed, what is the essence of the suggestion that centralized planning of these indices should be carried down only to the level of the economic council, and that below this level the enterprise itself should determine its wage fund, the level of labor productivity and capital investment? If centralized planning is brought down only to the economic council level, and the enterprises are permitted to plan these indices themselves, then what is the point of planning at all levels down to the economic council? With an artificial gap between planning for the national economy and for the enterprise, the economy will be "running idle." The mechanism suggested actually spells a weakening of unified national economic planning, the isolation of the enterprise from the common targets of the national economy. We cannot help but agree with Comrade

Sukharevskii, who noted that some economists do not understand the relationship between the reproduction process at an
individual enterprise and in the national economy as a whole,
between circulation at an enterprise and circulation of the
whole aggregate social product. Such a gap may create an
altogether abnormal situation.

If, for example, the enterprise planned the wage fund itself, as some economists and managers propose, this would
complicate unified planning and especially the implementation
of a unified wage policy in all branches of the national economy. This policy is connected with the distribution of labor
resources, which in this case might not correspond to the
interests of the national economy. The point is that if the wage
fund is not planned on a unified scale—for the national economy—but by the enterprises themselves, this would exclude
the planned distribution of manpower in the national economy.

We must also agree with what K. N. Plotnikov and A. G.
Zverev said concerning capital investment — this important
tool of planning the national economy — since it is capital investment that predetermines economic proportions for a long
period of time. Even in capitalist France, where "indicative
regulation" is practiced, there is the effort to exert influence
on the enterprise through investment. But in our country, in a
socialist society, it is suggested that we permit enterprises
to plan their capital investment outside of a unified plan. This
would actually mean giving up a unified capital investment
policy, the planning of structural changes and proportions in
the national economy as a whole.

Nor is it correct, in our opinion, to replace the system of
material-technical supply by trade in the means of production.
In a socialist economy such a method would be unacceptable.
It is doubtful that the distribution of material resources through
mutual trade relations between enterprises will correspond to
precisely those proportions which are essential to attain the
targets embodied in the national economic plan.

Here I should like to say a few words about what enterprises actually need. Not so long ago I was in Orenburg and had
talks with some staff members of the economic council and
of enterprises. They have come up against two difficulties in
planning: the first is that there is no stable plan — assignments
and product assortment are frequently changed; the other difficulty is that enterprises are not given long-term plans. It is
high time to begin long-term planning at enterprises.

It is essential that we work out suggestions concerning long-term planning at enterprises. A system of continuous planning must be set up so as to have an annual plan, target figures for the following year, and a long-term plan. Within this long-term plan there can be a bonus scale of a long-term nature, which would be substantiated by this plan but not necessarily based on profitability. In some cases the bonus scale may be based on profitability, in other cases on economies from reduced production costs, in still other cases on economies per ruble of marketable output, etc. It is essential that these scales be differentiated, of a long-term nature, and, what is most important, substantiated by the state long-term plan. All this must be part and parcel of the system of long-term planning. This will create the possibility of organically combining the unified planning of the national economy with a system of methods of economic incentives.

The 1964-65 Discussion

Kommunist, 1964, No. 5

Socialist Economic Management and Production Planning

V. S. Nemchinov

In the course of our socialist construction we have gained vast experience in planning the national economy and guiding the processes of economic development. Soviet economists are now confronted with the urgent task of drawing theoretical generalizations from this experience of nearly half a century.

Special attention should be given, first of all, to a comprehensive study of the pattern of relations between individual economic management units and the national economy as a whole, which arise in the process of planning and managing the sphere of material social production. As time goes on, this problem acquires greater urgency because our plans are becoming increasingly comprehensive and all-embracing. The principle of planning now penetrates literally every pore of the national economy. Hence, the extension and deepening of the role of conscious planning in the socialist economy demand that the techniques and methods of planning conform more closely to the principles and requirements of the vital, immediate, and continuous economic process.

1

In the process of drafting and implementing a plan, decisive importance should be attached to the proper distribution of functions, rights and duties between the planning bodies and executive economic links. No superior agency can have as good a knowledge of internal resources and conditions of production as the enterprise itself. Under centralized planning, therefore, it is especially important to observe the necessary measure of centralization, so as to be able always to ensure adequate scope

173

for local planning and initiative. No increase in the number of plan indices can ensure rational economic interrelations between the center and the localities, between the central planning bodies and the enterprises.

We know that national economic planning is merely one aspect of the management of social production. To be specific, one can have a good plan but do a poor job of preparing for its implementation. It is obvious that the conversion of intermediate estimated indices into guiding plan indices, which are designed to reflect only the ultimate result of socialist economic management, is an incorrect approach. Such an extension of plan indices can only result in the economic system beginning to lose the necessary flexibility and operativeness. Such a situation leaves too little scope for the local creative initiative of the primary production units and individual economic executives. Such an incorrect extension of the planning principle is bound to give rise to a system of bureaucratic administration that is based on the incorrect notion that planning personnel and administrators alone can direct social production, without mobilizing creative managerial independence, without utilizing the principle of giving the working people a personal material interest in the results of collective labor.

It is also known that some plan indices (gross output, for instance), while successfully performing their function at the level of national economic planning, at the same time manifestly distort the results of the economic activity of individual enterprises. On the basis of this contradiction, for instance, there arose the problem of improving certain plan indices. In particular, it has been proposed to introduce a special plan index — "normative cost of processing." Experience has shown that in many branches this index fairly adequately reflects the amount of work carried out by the personnel of an enterprise. Quite often a situation arises in which the aggregate result of joint coordinated efforts in no way corresponds to the sum total of individual results. It is possible, for example, to have an increase in per hectare yields at each individual collective or state farm for the current year, while the average yield for the economic region as a whole registers a decline, if the increase in crop areas in individual parts of the region obtaining a lower yield is bigger than the growth of crop areas in other parts of the region obtaining higher yields.

It would be wrong to believe that the chief difficulty in the way of extending and deepening the principle of planning in our economy and converting it into a comprehensive and all-embracing principle consists in the choice of a definite system of summarized (uniform from top to bottom) and differentiated plan indices. This gives rise to a more complicated problem, as was at once revealed when our planning bodies began to implement the directive stipulating that our national economic plan be based on plans drawn up directly by the enterprises. It has become obvious that this correct directive must not be understood in a primitive way, that is, as a demand to make the national economic plan a mere sum total of the local plans. The national economy is a complex economic system, and it is not identical with a simple sum total of its elements and primary cells. It consists of a number of smaller economic systems (for example, branches of production, economic areas, economic associations). These smaller economic systems, in turn, are made up of primary production (enterprises) and consumption (families) economic cells. Direct as well as reverse connections always arise among them. Of quite essential significance in this complex system of relations are various stimulators, notably material and moral incentives. It is impossible to measure the entire complex of such relations by merely summing up planned and report indices. Even the ordinary information contained in the local plans cannot be merely summed up. It must be qualitatively and quantitatively transformed before it is utilized for the drafting of a plan in a higher agency.

It is well known that the consolidated plan technological norms, being average for the association as a whole, largely depend on the planned assortment of the produce turned out by the entire association, the distribution of the production program among individual plants, the technological methods of production, and the raw materials used. Consequently, the consolidated system of norms, for example, of material supplies for the entire association, can be obtained only in the form of norms that are transformed accordingly, i.e., weighed according to the factors on which the level of primary technologically substantiated norms depends. The same is true of the machine-time norms in planning the exploitation of machine tools and production equipment, as well as of a number of other similar cases.

That explains why the consolidated norms can never be permanent and immutable. Only when we have a more perfect system for processing them on electronic computers will it be possible to ensure the necessary flexibility and changeability of summarized consolidated norms so that they can adequately reflect the composition of the primary technological norms comprising them.

When smaller systems are merged into large economic systems, the structure of the bigger system is always of decisive significance. And this circumstance is completely ignored when a bigger economic system is viewed in a primitive way, as a mere sum total of the primary smaller economic systems.

A primitive understanding of interrelations between big and small economic systems can only create an ossified mechanical system in which all control parameters are given beforehand and the entire system is limited from top downwards for each given moment and at every given point. Life is bound to introduce very substantial correctives into such a system, with the result that the plan indices will not have the necessary definiteness and will become "elastic." Such an economic system, limited as it is from top downwards, will act as a brake on social and technical progress, and sooner or later it will be swept away under the pressure of the real process of economic life.

For economic systems of varying degrees of complexity to exist and operate harmoniously, it is not at all necessary to strive for their mechanical and arithmetical identity; it is quite sufficient to ensure the necessary priority for a bigger economic system and such a transformation of the flow of economic information that will guarantee uninterrupted operation of the mechanism of reverse economic ties (reactions correcting deviations from the program).

These reverse ties in the economy can be realized only through the system of cost accounting and the system of diverse public funds, notably the enterprise funds (for instance, the material incentives fund, the enterprise expansion fund, the new technology fund, etc.). The procedure for replenishing and allocating these funds should be regulated by long-term norms fixed by economic legislation. The main thing is to combine the mechanism of planning with the system of cost accounting and the system of enterprise public funds. It is important to find, as soon as possible, ways and means of establishing a social mechanism of this kind. This is the key to the solution of the problem of rational planning and management of material production. And only in this way will it be possible to set in motion the mechanism of reverse ties.

The mechanism of reverse ties should be adjusted before-

hand in such a way as to make the system of cost accounting
and the funds system stimulate complete fulfillment by economic
associations and enterprises of the targets and orders fixed in
the national economic plan. The long-term norms, which regulate the system of enterprise public funds and the system of
cost accounting relations, must be so chosen and corrected as
to continually secure and maintain the economic proportions
which conform to the basic directives of the national economic
plan. The entire system of economic levers (planned prices,
credit, subsidies and grants, financial sanctions) will be used
in the same direction.

2

Cost accounting relations should serve as the basis for bringing the economic potentialities of enterprises into accord with
the requirements of the entire national economy. This can be
achieved only by radically altering the pattern of relations between the planning bodies and the local enterprises. And this
requires, in particular, that the planning bodies distribute, in
an economically effective and rational way, sufficiently profitable orders—based on the national economic plan—among enterprises and construction sites through the network of economic
agencies. Each enterprise should submit to the planning bodies
preliminary proposals concerning the conditions under which it
is prepared to fulfill one or another plan order for the delivery
of goods, specifying the assortment, quality, time limits, and
prices. The economic and planning bodies, for their part, should
distribute their orders only among those enterprises whose
terms for carrying out plan orders are most advantageous for
the national economy as a whole.

The agreement of an enterprise to accept a definite plan
assignment, being confirmed by a written document, converts
the plan assignment into a plan order. As far as the planning
bodies are concerned, this procedure is more complicated, but
it is necessary as a filter against manifestations of pure voluntarism, and it is quite feasible. This system can be called a
cost accounting system of planning because it organically combines the planning and cost accounting principles — the principles which should regulate any type of economic activity in
conditions of socialism. What elements comprise this system
and what conditions ensure its uninterrupted operation? The
cost accounting system of planning will operate smoothly only

if it is based on the following principle: everything that is useful and advantageous for the national economy as a whole should also be advantageous for the enterprise carrying out the corresponding portion of the plan as an executive link. The operation of this principle can be guaranteed if the plan assignment is transformed into a plan order and if the basic conditions for fulfilling the order are established, in particular, if the price is acceptable both to the planning body and the enterprise.

The contractual relations arising under this system among enterprises, economic associations, and planning bodies cannot be limited to a term of one year; they must operate over a longer period. Under stringent annual calendar planning, the continuity of the economic process inevitably comes into sharp contradiction with the intermittent, discrete character of the planning process. If the entire process of socialist economic management were to be confined within a narrow calendar framework, it would be tantamount to planning the economy anew every year. The living tissue of the economic process is inevitably rent and, as a consequence, the inner continuity and consistency of the very process of economic management are violated.

And yet each calendar segment of the production process is inseparably and closely linked with the preceding periods, constituting a single, multistage process of economic development. The new requirements arising in this process must be formalized by supplementary contractual relations. Hence, the whole business can be reduced only to the distribution of new orders and, whenever necessary, to the correction of old ones. Indeed, the portfolio of orders will change in any calendar period, which is fully in keeping with genuine continuity of the economic process!

However, the economic contracts must stipulate definite obligations not only for local enterprises, but also for higher economic and planning agencies. For the planning organizations, these obligations are expressed in the price fixed by the contract and in the obligation to buy up the entire ordered assortment of goods through precisely defined economic agencies.

The abnormality of the existing planning procedure consists in the one-sided character of the obligations. Our local enterprises are constantly getting from above definite planned percentage assignments (concerning growth of the volume of production, higher labor productivity, lower production costs, etc.), while higher bodies, as a rule, do not bear any responsibility to

local enterprises for disproportions in the plans. Not infrequently, the plans dealing with production, labor, finance, credit, and material and technical supply are uncoordinated. The
reason for this is that individual elements of the economy are
planned separately. The mechanism of present-day planning is
so constructed that each line and column of the plan has its
own master, while the integration of the plan is not ensured
organizationally. Changes in some of the plan indices are by no
means always accompanied by corresponding changes in other
indices. Mutual obligations, on the other hand, are not always
formalized by contracts with customers and suppliers. Contract
discipline is very weak.

Only the conversion of the plan assignment into a plan order
and the corresponding formalization of the plan order, on a
cost accounting basis, through concrete economic associations
and enterprises will make it possible to dovetail the individual
lines and columns of the national economic plan.

The cost accounting system also eliminates another essential defect of current planning procedure — the separation of
price planning from the planning of volumetric quantitative indices. Meanwhile, the national economic plan can fully conform
to the law of proportional development if the prices envisaged by
the plan fully correspond to the level of labor, material, and
monetary outlays and if they guarantee the minimum level of
profitability to every enterprise that is operating normally.

3

The national economic plan can be coordinated in all its parts
and produce an optimum effect only if it is based on a system
of prices which fully corresponds to the volumetric indices of
the plan in the sense of balancing production and consumption
and ensuring economically effective employment of labor, production and natural resources.

A powerful, large-scale and diversified socialist economy
can in no way be regarded as a natural economy based on direct
natural exchange of products. Even in the process of transition
to communism, society cannot but compare the labor outlays of
its members and the results obtained, cannot get along without
assessing the results of the economic activity of its primary
production-consumer units, cannot but take into account the
economic effectiveness of the use of production capacities.
Society cannot develop on a planned basis without determining

the extent to which economic and natural resources are used rationally.

Under socialism, money and commodities no longer serve as a means of appropriating the surplus product and exacting a toll from the working people, as is the case under capitalism. But money and commodities continue to perform very important functions even in a socialist society. In extending the social (branch and territorial) division of labor, decisive importance is acquired by the production of commodities designed not for personal consumption but for use by other units of society.

A permanent shortage of particular material and technical resources is primarily determined by the fact that we continue to regard articles of supply not as commodities (with their laws of equivalent exchange) but as objects of direct product exchange. To ensure continuity of the process of equivalent exchange, the items manufactured by state enterprises must be exchanged, in the main, through the system of wholesale state trade. In essence, however, a peculiar "system of rationing" continues to operate in the sphere of material and technical supply. The distribution of articles of material supply is based on orders issued from previously allocated physical funds for each individual enterprise, with every item detailed. All material and technical resources are scattered among innumerable individual fund-holders (organizations and enterprises) and individual detailed commodity positions. The cumbersome system of preliminary orders for funds with subsequent repeated reexamination of funds, followed by stringent formalization through the system of orders for delivering material supply items to consumers, inevitably leads to metabolic diseases in our economic organism. Like any ration card, the order always specifies the delivery of the entire quantity of goods indicated in the funds and orders and not of the amount really needed at the given time and place. As a result, material values that are not needed at the given moment accumulate in some links of the national economic organism, while other links experience an acute shortage of these values.

In the overwhelming majority of cases, a shortage of material resources is attributable to such an irrational system of their distribution. All economic units will ultimately be provided with the necessary materials, but the imperfect system of their distribution leads, on the one hand, to the freezing of circulating funds, and, on the other, to the constant shortage of articles of material and technical supply.

It should also be emphasized that the process of production and circulation can proceed without hindrance only if the national economic plan is based on a joint system of equations in which economic estimates and plan prices are reciprocally coordinated and correspond to the physical (volume) structure of production and consumption. In conditions of the capitalist market economy, this process proceeds spontaneously, behind the back of the producer; in our country it must be systematically and consciously regulated by a corresponding system of commodity exchange.

In socialist society the acute need for a balanced and mutually coordinated system of plan prices and volumetric plan indices also has the force of an immutable law, though the very process of mutual coordination and balancing must be effected consciously by tracing the behavior of the given system (the material supply system, for example) on electronic computing machines.

The regulating power of prices is so great that bourgeois economists usually oppose prices to the plan and often proclaim the slogan of "price instead of plan." We Soviet economists, being confident of the regulating power of the plan, sometimes arrive at the opposite assertion: "plan instead of price." Meanwhile, in our view, the only correct solution of this question consists in purposefully combining plan and prices. The priority of the plan in this matter is expressed in the fact that the prices themselves are also subject to planning; the plan prices must conform in full measure to the objective process of price formation and correspond to the processes of creating and redistributing value.

Planned control over the conformity of prices to their level of value can be effected in the form of definite and rigorous economic and plan calculations, which will reflect not only the estimating, but also the stimulating and redistributing role of prices. This will serve as the basis for constant and systematic control of price proportions and their conformity to the level of value.

Yet the very procedure of current planning of prices can be organized without the present cumbersome system of planning, which includes centralized compilation of detailed price lists. In the first place, stable, rarely changing prices should be fixed only on a limited number of particularly important commodities. This category includes the most essential commodities determining the material and cultural standard of the population, as well as commodities determining the level of production costs. Stable prices are necessary only for basic serially manufactured commodities; for the other commodities belonging to this group, strictly limited correlation of prices can be established in relation to the initial basic commodities. To bring these prices into conformity with those arising from the balance of production and consumption and the balance of supply and demand, it would be expedient to make use of a special national price-regulation fund. Most of the other commodities can be put in a group for which only so-called "controlled prices" are approved. These prices are worked out by economic associations and agencies and endorsed by the Supreme Economic Council of the USSR or by the governments of the union republics. On the other hand, the prices on all commodities that are not produced serially can be fixed by the enterprises themselves on condition that they adhere strictly to the approved method for calculating such prices.

A more flexible and improved procedure of price planning will make it possible to rule out completely the practice of contraposing the plan to prices and prices to plan. Only in this case can prices perform their chief role of an economic regulator of the exchange of the results of labor among individual economic units of society. Only under this procedure will the planned prices regulate the exchange of some component elements of the social product for others. Such a system is fully capable of maintaining the required proportions and systematic development of the entire economic system.

The systematic use of prices in the process of national economic planning fully corresponds to the principles of democratic centralism. In the form of stable and, at the same time, flexible prices, all the economic units of society will be provided with a reliable criterion for choosing the optimal regime of economic activity, under which the local (private) optimum will be combined in full measure with the general (national economic) optimum.

4

Under the cost accounting system of planning, the principle of the profitability of the economic operation of an enterprise through plan prices and the system of public funds (whose replenishment and expenditure are regulated by legislatively instituted long-term norms) is dovetailed with the general planning principle of proportional economic development of the entire economic entity. Under this system of planning, ·society as a whole will appropriate the basic share of the profits accumulated in the process of production, while at the same time the economic units themselves will acquire the necessary material interest in raising the profitability of their economic activity. It must be borne in mind that the material interest of the given production collective in the results of its economic activity is chiefly determined by the share of the profit contributed to diverse public funds of the enterprise (the fund of additional collective labor remuneration, the enterprise expansion fund, etc.).

In order strictly to define relations between financial bodies and the enterprises, it is necessary, by legislation, to establish long-term norms which regulate in advance the sufficiently big part of the actual profit that is placed at the disposal of the enterprises. Moreover, if the enterprise manufactures commodities which, for one reason or another, fail to meet the demand for them, the contributions to the enterprise fund can reach the maximum level, and, on the contrary, if the manufactured commodities are not in demand, these contributions can be reduced to the minimum, even to naught.

Since in the final analysis the planned economy can have only one — state — pocket, the opinion prevailed at one time that there is no sense in legislatively regulating the processes of profit distribution between the state and the enterprise. However, experience teaches that even in one pocket it is extremely important to distinguish between separate purses. Only if this is done is it possible to ensure a differentiated approach in regulating the regime of socialist economic management and the course of the economic processes themselves.

Society has a clearly expressed cellular structure (family household economy, enterprise, association, the state). Hence, one cannot be indifferent to the way in which the relations between individual cells of society, as well as between each cell and the entire national economy, are regulated.

In the theory of economic management, one must distinguish administrative and executive links, and one must single out in particular the parameters of management among the characteristics of one or another economic system. For the capitalist economy, taxation rates, interest rates, subsidies, and grants function as the parameters of management. In the planned socialist economy, the control figures determining the plan assignment for the executive links of the system are paramount among the parameters of management. However, even under socialism the financial and credit levers also have a definite importance. In these conditions it is extremely important to refine the methods and technique of planning with a view to utilizing the economic levers which stimulate the economic activity of the enterprises in a direction determined by the national economic plan.

The current practice of determining plan assignments in percentages of the attained level puts our advanced units in an unfavorable position. Under this system, higher targets are constantly imposed on the advanced enterprises, while the lagging links of the system are given lower and easier targets.

Even if the new plan assignment is raised by the same percentage, the advanced enterprise will be placed in less favorable conditions because the absolute size of one percent of plan increase is much greater in the case of the advanced enterprise than in that of the lagging one. Under the existing system of percentage planning based on the attained level of production, the efforts of the collective are often nullified by the subsequent plan, for it at once absorbs the entire result of the collective's previous efforts. Therefore, the heads of enterprises are often compelled to conceal their resources and reserves from the planning bodies and to operate below capacity in order to avoid putting their collectives in the difficult position of not being able to attain the higher targets fixed in the next planning period.

The system of planning based on the attained level of production creates completely abnormal interrelations between the administrative (in particular, planning) and executive links: the latter try, by every possible means, to secure a minimum plan target for output and a maximum target for capital investments, while the administrative links strive for the opposite result. It is no accident that our planning and statistical agencies have a very superficial knowledge of the production capacities of enterprises and industrial branches. Yet national economic planning must be based precisely on a systematic expansion of production potential, which is effected not only by investing capital in new construction, but also by raising to the utmost the volume of output per unit of available production capacity.

5

The paramount task in drawing up the technical, production and financial plan is to arrive at a correct estimate of the production capacities of the national economy. The production capacities depend on the assortment of planned output, the quality of fuel and raw materials, and the possibility of eliminating "bottlenecks" that limit production. Therefore, it is desirable that the local technical, production and financial plans should always contain several variants for increasing production capacity as a result of changing the planned assortment of goods, changing the quality of fuel and raw materials, approving allocations for capital investments needed to eliminate "bottlenecks," etc. All the variants should be accompanied by estimates of factory production costs, labor productivity, and capital investments per unit of output.

Under such conditions the task of the planning bodies and economic associations will consist of correctly distributing the plan assignment-order, making maximum use of the production capacities of enterprises. The main task at present is to combine in full measure the fulfillment of the national economic plan with economically effective utilization of such resources as production capacities and land. The resolution of the CPSU Central Committee and USSR Council of Ministers of March 20, 1964, on the practice of planning collective-farm and state-farm production indicates the correct way of solving this problem. The substance of the matter lies precisely in combining state leadership with the all-round development of creative initiative and activity at the local level. This applies in equal measure to all branches of the economy. The cost accounting system of planning will permit the proper utilization of the production capacities of industrial and construction enterprises.

Each production collective is provided by society with certain production capacities. By virtue of this, each production collective is obligated to ensure their profitable operation. Hence, the full factory cost of output must comprise not only the factory production cost, but also the compulsory normative charges on the fixed and circulating assets placed at the disposal of the enterprise by society. Such charges will make it incumbent upon the enterprise to ensure the minimum economic effectiveness (profitability) of the assets functioning in its production.

Intensive research work is now being done to elaborate objective methods of determining profitability norms for the fixed production assets and to estimate the economic effectiveness of capital investments. The results obtained so far justify the assumption that methods will soon be evolved for determining the norms of compulsory charges on fixed assets; these norms will take into account both the material composition of the assets and the economic effectiveness of their utilization in different branches of production.

The immutable law of economic development, formulated in the CPSU Program, envisages a maximum economic effect with a minimum expenditure of labor, materials, and money.

Normative charges on fixed and circulating assets represent one of the forms of expressing the minimum demands of society in relation to the expected economic results from the use of production resources. This is one of the economic criteria of the optimization of the technical, production and financial plans drawn up by enterprises. Charges on fixed assets should be regarded as an objective method of planning the profitability of the enterprise. This method is substantially better than the present method of planning the profitability in percentages of the factory production cost. The latter is illogical, for a rise in the cost of production should be accompanied by a decline in profitability, and not an increase. Moreover, the introduction of differentiated compulsory charges on fixed assets makes it possible to take proper account of the economic conditions of operation of individual enterprises, with their differing amounts of assets per unit of labor and, consequently, their differing levels of labor productivity.

The introduction of compulsory normative charges on fixed assets will make it possible to put an end to the latter's gratuitous character, thus making for more economical utilization of production assets, bringing down the capital requirements per unit of gross output, and reducing the capital requirements per unit of production increase.

The planning of production capacities, as well as the planning of profitability in the form of charges on fixed assets, are indispensable elements of the cost accounting system of planning.

The transition from "plan-assignments" to "plan-orders," just as orientation toward the planning of production capacities, will do a great deal to eliminate bureaucratic administration and the various manifestations of voluntarism in the sphere of economic management and planning.

The cost accounting system of planning creates a reliable filter against the survivals of voluntarism in the links of social-ist economic management. It can be asserted that even in our conditions elements of voluntarism, should they be allowed to develop, can lead to consequences which, in individual cases, will be no less harmful than the results of spontaneous com-petition under capitalism. In conditions of administrative vol-untarism, the creative freedom of economic managers in choos-ing ways and methods of fulfilling the plan is extremely limited. The situation here is analogous to that of the steam engine at the dawn of the industrial revolution. For instance, at the time of Newcomen, prior to the invention of Watt's governor, the movement of steam in the engine was regulated by a special man who pulled the handle of the slide valve at the proper moment.

Just as in regulating machines in mechanics, such a situa-tion cannot be tolerated in managing the economy. In conditions of modern serial, automated technology, we cannot reconcile ourselves to such an obsolete method of production manage-ment.

6

In the process of transition to the cost accounting system of planning, there arises the acute need to introduce automated regulation systems in the national economy. This implies, first and foremost, an automated system of collecting, processing and transforming economic information (report and plan), as well as an automated system of economic and plan calculations as a basis for plan decisions arrived at by central and local agencies.

At the present time, in accordance with a decision adopted by the CPSU Central Committee and the Council of Ministers of the USSR, extensive work is under way to design a uniform automated system for collecting, transmitting and processing economic information, to establish a network of dispatching and computing centers, and to elaborate appropriate mathematical-economic and mathematical-statistical models of economic pro-cesses and national economic planning. The training of spe-cialists in this field is also under way.

The automated processing of flows of economic information, going from the bottom up, will ensure current supervision over the fulfillment of the national economic plan and, whenever necessary, the timely regrouping of elements in the national economic plan so as to maintain proportionality in the entire process of economic development. Inasmuch as such automated electronic regulation systems are based on cognized necessity, they cannot be identified with spontaneously operating processes. Only by the extensive use of an automated system of economic planning and management, based on modern electronic techniques, can we banish spontaneous processes from social life.

Automated electronic regulation systems are called upon to ensure priority for the directives and control figures of the national economic plan and, simultaneously, to promote the extensive use of cost accounting and economic levers in the form of the system of public funds, prices, profits, and credits. The existence of an automated system for the collection and processing of economic (report and plan) information will make it possible to make wide use of the system of public funds.

Moreover, extensive use can also be made of the long-term norms fixed by economic legislation. For example, these norms can regulate the size of contributions from enterprise profits to the state budget, as well as to the material incentives fund, the enterprise expansion fund, etc. The contribution quotas, in particular, may depend on the relationship between the actual and the normatively estimated profit (calculated as a percentage of the value of the enterprise's fixed and circulating assets). It is expedient in this case to fix sales prices in such a way as to ensure that the share of the profit contributed to the state budget should at least be average for the given industry.

Under the cost accounting system of planning, economic legislation should provide differentiated quotas for distributing profits between the state and the enterprise, making special provision for cases in which the actual profits do not exceed the normatively estimated ones. If the actual profits are below the normatively estimated ones, then the enterprise funds might obtain a definite share of the savings derived from bringing down production costs, cutting expenditures from the wage fund, and increasing labor productivity. This share can also be fixed legislatively in the form of long-term norms.

7

The entire system of economic levers must be regulated through long-term norms and prices in such a way as to make it advantageous for enterprises to fulfill the directives and control figures of the national economic plan. And this task is quite feasible. We can no longer tolerate a situation in which particularly scarce or strategically important commodities bring a loss to enterprises, while commodities that are not in demand are highly profitable.

In the socialist economy, developing under the powerful influence of the law of planned (proportional) development, the law of value is also of essential importance. The cost accounting system of planning makes it possible organically to link the operation of both laws and, on this basis, constantly to combine the national economic optimum (taking shape under the influence of the law of planned, proportional development) with local optimums (arising under the influence of the law of value).

With the aid of a stable but flexible system of prices, a rational system of public funds, and a full-blooded system of cost accounting, it is quite possible to ensure harmonious operation of these two laws, for in that case the operation of the law of value will be subordinated to the operation of the law of planned (proportional) development of the national economy.

Under this system of planning, the correspondence of the overall and local optimums will be founded on a system of prices which, like economic multipliers, will not only balance the entire system of the national economic plan but simultaneously stimulate the economic activity of all enterprises toward fulfillment of all the plan targets.

Under a properly functioning system of prices and a system of legislatively fixed long-term norms, it is possible to create a situation in which the enterprise will strive to obtain orders from the planning bodies and economic associations precisely for those types and assortments of commodities which are envisaged in the overall national economic plan.

Such a combination of different optimums (general and particular) gives local enterprises and associations the necessary economic initiative and leeway for selecting ways and means of fulfilling a plan order. And for the national economy as a whole, it proves possible in the process of plan fulfillment to meet the requirements following from the national economic plan. The continuity of the economic process in this case is ensured by the fact that the annual current plans are drawn up as definite calendar sections of a long-range plan, while the annual and current calendar plans drawn up by the enterprises are determined by their portfolios of orders and the time limits for fulfilling these orders, as specified by the economic contracts.

Only the time limits for fulfilling the plan of deliveries, specified by the economic contracts and the portfolio of orders, can reflect in full measure the commodity requirements for calendar periods shorter than a year.

If the current annual national economic plan is drawn up well, there is no need to strive for mechanical and arithmetical identity between annual indices of the overall national economic plan and the annual dissection of the obligations assumed under the economic contracts. Then the national economic plan will be able to perform its proper role as a general guide for the economic activity of individual enterprises and their associations.

Moreover, a uniform system of economic information and its processing in electronic computers ensure current supervision over the degree of correspondence between indices of the current national economic plans and the calendar plans of enterprises based on fulfilling the portfolio of orders. The discrepancies in these indices, as revealed by economic analysis, will serve as a basis for corresponding instructions from the administrative links of the economic system, addressed both to the executive links (enterprises) and the planning bodies. In any case, the existence of substantial discrepancies between plan indices and the portfolio of orders may require not only adjustment of the economic contracts, but also adjustment of the national economic plan itself. The latter is especially important in drafting plans for subsequent calendar periods.

Let us sum up. The cost accounting system of planning, aimed at combining the procedure of national economic planning with the contractual system of cost accounting and the system of material incentives, consists of the following links and has the following structure:

1) The enterprises submit to the planning agencies of economic associations and economic councils variants of a plan to expand their production capacities, specifying operating and capital outlays. Proceeding from the national economic plan and these variant plan-programs, the planning agencies distribute among economic organizations and enterprises the plan-order, formalized in the shape of a contract indicating the price and other terms of delivery.

2) The enterprises calculate not only the production cost, but also the full factory cost, which is equivalent to the present factory production cost plus compulsory charges on fixed and

circulating assets, calculated on the basis of legislatively fixed long-term norms. The results of economic activity are defined as the difference between the planned price and the full factory cost. A positive difference characterizes additional profit, a negative difference — outstanding profit. The actual profit is defined as the sum total of compulsory charges on fixed and circulating assets, plus additional profit or minus outstanding profit. Compulsory contributions to the enterprise funds (separately to the material incentives fund and to the factory expansion and new technology funds) are made from the actual profit in accordance with long-term norms. Under certain conditions, contributions to enterprises are envisaged also from savings in production costs and the wage fund (compared with the preceding period, per unit of output).

3) The planned wholesale prices are fixed at the level of average branch cost. If there is lack of coordination between the systems of wholesale and retail prices, a fund for the regulation of prices is established. It is built up from national revenue and is expended on subsidies to the wholesale (marketing and procurement) trade network. Thus, for the procurement prices of food and fodder grain (whose cost is planned on the basis of average yields over a number of years), a special insurance fund is introduced. Premiums are issued to agricultural enterprises from this fund in order to guarantee the level of profit envisaged in the planning of zonal prices in cases when actual yields are below those used in calculating the plan price because of unfavorable weather conditions. A special system of plan calculations operates to exercise control over the correspondence of prices to the level of the socially necessary outlays of labor. The current centralized procedure for drawing up price lists is substantially simplified and decentralized.

4) Instead of the system of physical funds regulating the use of material resources, there comes into operation a ramified system of special-purpose public cost funds, which regulates the entire process of expanded reproduction both on the national-economic and enterprise level. The system of special-purpose cost funds embraces not only special cost funds of material resources (fuel and power, raw materials, the spare parts and depreciation funds), but also the labor remuneration, premium, material incentives, public consumption, and other funds. The replenishment and expenditure of this entire system of special-purpose public funds are regulated legislatively on the basis of long-term norms.

5) With the aim of radically improving the procedure and technique of planning and managing social production, it is intended to establish an automated electronic system for collecting and processing economic information, as well as an automated system for carrying out mass economic, planning, and optimal engineering-technical calculations required for the adoption of substantiated economic and planning decisions by the central and local agencies.

The cost accounting system of planning, based on formalizing the demands of the national economic plan through the system of economic contracts, undoubtedly has immense advantages over the existing system of percentage distribution of national economic plan indices among economic organizations and enterprises.

The cost accounting method of planning also fully conforms to the principles of democratic centralism. It will help eliminate bureaucratic methods in planning and in managing the economy.

There is no more important task now than to give wide scope to the vital uninterrupted process of economic activity by bringing the methods and techniques of production planning and management into accord with the continuity and flexibility of the vital process of direct socialist economic management.

Pravda, August 17, 1964

For Flexible Economic Management of Enterprises*

V. Trapeznikov

The decisions of the 22nd Party Congress and the speeches of N. S. Khrushchev repeatedly stressed the importance of material incentive based on the correct economic assessment of enterprises' activity.

To this day the assessment of an enterprise's efficiency is based on a multitude of standards which fail, however, to provide a general picture of its activity. The plans fix the norms for the wage fund, office outlays, warehouse reserves of material, expenditure of materials on the repairing of equipment, the number of administrative and management personnel, and many other things. An enterprise manager is extremely limited in his actions; he does not even have the right to determine the number of personnel in the different categories and to create the necessary subdivisions within the limits of the general wage fund. These norms and restrictions are, as a rule, established on the basis of the enterprise's activity in the preceding year, and not on the basis of actual need or scientific calculations.

Such norms were useful at a certain stage of our economic development. But now, with the increased efficiency in production and management, these norms have largely become

---

*From the editors: Academician Trapeznikov's article is published as a presentation of the problem, and the editors invite scholars, and workers in industry, construction, transport, and trade, and in party, economic, and planning agencies to express themselves in <u>Pravda</u> on the problems treated in this article.

The author is a Member of the Academy of Sciences of the USSR.

obsolete; they have turned into petty tutelage, binding the managers' activities, complicating accounting, and increasing the bookkeeping and control staff. And even with the help of the norms it is impossible to foresee all the circumstances that arise in practice. Can we, for instance, foresee that an inefficient manager of a dairy plant will daily discard several tons of whey, which could be used for fattening pigs, as <u>Pravda</u> recently reported?

Technical progress changes the nature of production daily, but the norms remain unchanged for a long time. For instance, the traditional accounting system includes tool shops, control-instrument and automation shops, and others as auxiliary subdivisions; norms are fixed for their staffs and they are subject to reduction. With the intensive mechanization and automation of production, it is clear that the reduction of these categories of workers can bring nothing but harm.

At most enterprises the administrative and management staff is cut annually by approximately five per cent, and it is invariably restored, in some way or other, in the course of the year to its former strength (if this were not so, it would have diminished to a half or a third long ago). Indeed, so long as the stream of paper resulting from extremely complicated accounting does not diminish, neither the volume of work in the sphere of supply nor the number of people handling the relevant documents can decrease. If the number of employees in these categories is, nevertheless, reduced, their functions are usually transferred to production personnel who, as a rule, are higher paid. Hence, what seems a saving is actually a waste of money.

And these numerous norms make it immensely difficult to assess the enterprise's activity! By concentrating the staff's attention on keeping to the fixed office outlays and other similar expenses, the norms divert its attention from the main issues of production efficiency, such as the quality of products and the general economic efficiency of the enterprise.

The time has come to discard the obsolete forms of economic management based on directive norms, and to pass over to a simpler, cheaper and more efficient type of control of the activities of enterprises. This control must be patterned so that the personnel of an enterprise find it economically profitable to organize their work along lines that are profitable to the national economy as well. Economic influence, in our opinion, must rely on a system of incentive, taxes, fines, and a flexible price system promoting the development of technology and the economy in the desired direction.

Economic influence is successfully employed even in capitalist countries. For instance, the U.S. Bureau of Standards worked out a tax on the "volume of sealed vacuum" in electron tubes. This compelled the firms operating in this field to perform research which has led to a sharp decrease of electronic device dimensions and to a considerable advance in electronics. Here is another example. U.S. authorities are aware that rapid progress in science is necessary. To stimulate this progress, they have abolished taxation of that part of incomes which the companies invest in research work (including subsidies to scientific-technical societies). This has promoted the growth of company laboratories, the development of scientific-technical societies and, consequently, technical progress.

In our conditions, when all the financial and economic levers are in the hands of the state, measures of economic influence will prove still more effective. However, for this purpose the financial agencies must change their approach to the system of taxes and fines: they must seek not only to collect money for the budget, but, first and foremost, to stimulate the economically expedient activities of enterprises. Budget revenues will then reach the highest level, too.

Economic control must rest upon simple but exhaustive forms of assessing the enterprises' activities. The methods for assessing economic efficiency trouble not only executives and economists, but also specialists in automation. What economic criteria should be fed into computers in order to find the optimum regimes for running enterprises? If different initial data are fed into the computer, the answers also will be different. For instance, if the machine is consulted on the actions most desirable for the national economy as a whole, it will give one answer. But the machine will give an altogether different answer if it operates with criteria which serve to estimate the work of an enterprise. A worker in the financial agencies will get still another answer. This stems from the lack of coordination of economic criteria (assessments) in the different parts of the national economy. It is clear, however, that the system of criteria must be patterned so that the machine gives the same answer in all cases. In other words, the economic interests of an enterprise must coincide with those of the national economy. Then our economy will develop at the fastest possible rate.

We regret to say, however, that the economic interests of the producer and the consumer (the national economy) quite often prove to be opposed. For instance, the manufacturing plant is not interested in improving the quality of its produce if this entails a rise in its production costs, whereas the consumer is interested in goods of higher quality even if they are somewhat more expensive.

The national economy is interested in the production of new types of goods, while the producing organization is not, because this adversely affects its economic indices and, besides, involves additional trouble. A manufacturing enterprise is interested in piling up large stocks of materials (to avoid stoppage resulting from supply failures), whereas the national economy is interested in reducing the freezing of assets and materials that are required by other enterprises. A building organization is interested in the quickest possible expenditure of expensive materials (laying cables and pipes, and installing equipment long before these units are put into operation), since this adds to the fulfillment of its plan in terms of money, whereas the national economy is interested in having these materials used as late as possible, closer to the instant when they are really needed, so as not to immobilize capital, and so on.

To exert effective economic influence on economic activity, it is essential to choose a criterion that characterizes to the greatest degree the operation of the enterprise and meets the interests of both the national economy and the personnel of the enterprise. We consider that, along with various other quantitative and qualitative indices, it is profit that constitutes such a criterion. This matter has been repeatedly discussed in the press. Indeed, increased output (plan) as well as lower production costs, which reflect the growth of labor productivity, are to the advantage of the national economy and increase profits. But they are also to the advantage of the personnel if profit is taken as the basis of moral encouragement and material incentive. By changing the prices at which the plant sells its produce, i.e., by changing the size of the profit, an increase in the output of the most important products may be economically stimulated, which will provide a powerful lever for managing the enterprise. The Party Program points out that the price system should be constantly improved, that "prices must, to an ever growing extent, reflect the socially necessary outlays of labor and ensure the recoupment of production and circulation expenditures and a certain profit for each normally operating enterprise."

The choice of profit as the basic criterion provides the opportunity to simplify considerably the regulation of an enterprise's activity. This basis makes it possible to cut substantially the number of expenditure items subject to rigid restrictions, since any unwarranted expenses will affect production costs and will be controlled through the overall index — profit. In reducing the number of standardized items, the special place of the wage fund must be taken into consideration, though even it need not be rigidly restricted. The enterprise may find it profitable to overdraw somewhat on the wage fund while achieving considerable savings on other expense items, on materials, for example. The point is that a saving of materials at a given plant is tantamount, in the final analysis, to a saving on wages in the national economy.

This is particularly evident as regards industries where wages are but a small fraction of the total cost of the product. A small overexpenditure of the wage fund will sometimes yield a tenfold saving in materials costs. In extreme cases, economic sanctions may be applied; for instance, in determining the profit (and, consequently, bonuses) it can be held that an overexpenditure of the wage fund by one per cent is equivalent to raising production costs by a certain percentage.

While good work should be encouraged, economic sanctions should be applied in cases of bad work, such as fines for delivery delays. The fine should be paid to the party that has suffered and must affect materially the size of profit and thus the size of bonuses to the personnel. One may say that the plant is not always at fault in failing to make deliveries on time, as it is sometimes let down by its suppliers. That is true, but if the system of economic influence is introduced throughout, all parties will keep to the terms fixed.

This is also borne out by experience in foreign lands. Foreign firms pay considerable fines for delays in deliveries. Moreover, they may lose their customers and go bankrupt. The economic pressure which the firms experience is so effective that delivery terms are adhered to with exceptional precision. Economic influence will also promote shorter delivery terms in our conditions.

Considerable economic influence may be provided by introducing price variability. This will make it possible to solve at least three problems. The first is to encourage the output of new goods. For this purpose, higher sales prices should be fixed on new goods so as to compensate their higher initial production costs and guarantee a profit to the plant and bonuses to its personnel. The price should be somewhat reduced each year in conformity with the reduction of production costs in the preceding year. (The plant will be interested in cutting production costs, because this will raise its profits and, consequently, the bonuses to its staff.)

Given such an organization of the matter, the enterprise will be economically interested in producing new items, and the national economy will be receiving increasingly cheap new goods. The high initial cost of the goods will cause no harm inasmuch as the absolute output of the first batches of such goods is not very large. The only aspect that may prove to be unusual is the report on the plant's operation: because of the reduction of prices, the rate at which the "gross output in rubles" grows may prove to be lower than planned, although the plant's actual production will grow as it should. This example goes to show once more that the index "gross output in rubles" frequently contradicts the economic interests of the country.

Plants are known to be extremely reluctant to raise the quality of their produce if this involves changes in production processes and higher production costs, despite the fact that such improvement is economically profitable to the consumer. To overcome this difficulty, it would be advisable to permit an increase in the sales price of the article if its quality is improved (a markup for quality), with the opportunity of reducing it later as production costs decline. In this way price variability will solve another problem: it will give the enterprise an economic interest in raising the quality of its product.

Finally, the possibility of fixing higher prices and, consequently, higher profits for new goods will encourage the priority output of the goods that are most important for the national economy. The state committees of the relevant industries must take part in the fixing of prices. In this way they will obtain economic levers for conducting an adequate technical policy. Consequently, price variability will also solve the third problem: it gives an economic stimulus to the most important technical trends.

Faster assets (capital) turnover, i.e., a shorter period of capital immobilization (surplus of materials in shortage, delay in use of materials in the process of production, lengthy periods of time for construction and the commissioning of production facilities), is of major importance in increasing the rates of economic development. This problem ought to be solved with the help of economic influence as well. Such influence must, evidently, be proportional both to the size of the immobilized capital and to the time that it remains immobilized. Consequently, the establishment of an interest rate on capital, and primarily on the circulating assets granted to the enterprise, will prove to be the best and, probably, the only form of economic influence. We often hear that an interest rate on capital is a category of capitalist society. This is not convincing. Indeed, the form is the same, but the essence is different. In capitalist society, capital is a source of income, and the interest rate is the capitalist's profit. In our country the interest rate received is not an income of the state (everything goes in and out of one and the same pocket — the state's), but a form of economic influence that speeds up the turnover of capital.

Thus, surplus materials in storage or delay in the use of materials in production, which immobilize circulating assets, will increase production costs, cut profits and, consequently, bonuses, because the enterprise will have to pay more interest. All this will compel the enterprise executives to reduce excesses.

How enterprises utilize their fixed assets (buildings, equipment) is another important factor. It is not being considered today in assessing an enterprise's efficiency. Meanwhile, many enterprises seek to obtain new capital investments while their fixed assets are used inadequately. The introduction of another norm would not produce any results. The only correct way to deal with this problem is through economic influence, for instance, by considerably increasing depreciation allowances, a portion of which must be paid back to the state (interest on capital), while another portion remains with the enterprise. (At present depreciation allowances form but a small fraction of the cost of the product and thus do not serve to encourage better utilization of fixed assets.)

The general aim of the above proposals is to substitute a sum of economic influences, which would channel the enterprise's activity, for control of each step of the enterprise executive through directives. Managers must be given extensive financial powers; they must be released from petty tutelage, and they must be obliged to ensure efficient operation of the plant and high quality of products. Meanwhile, the chief criterion (profit) will make them reduce all excesses and look for every possible way to reduce the cost of the goods produced.

Of course, the elaboration of measures of economic influence should proceed from the principle of maximum development of the democratic foundations of management coupled with a strengthening and improvement of centralized economic management by the state. This involves the further improvement of our economic planning and management, as well as implementation of the targets put forward by the Party Program: "Extension of the operative independence and initiative of enterprises on the basis of the state-plan targets is essential in order to mobilize untapped internal resources and to make more effective use of capital investments, production assets, and financial resources. It is necessary to enhance the role and interest of enterprises in introducing the latest machinery and using production capacities to the utmost."

Economic control ought to be introduced in many fields of the national economy, including trade. The cost of commodity distribution, i.e., the cost of the goods transfer from producer to consumer, should serve as the measure of the trade network's economic efficiency. This cost must include all the losses sustained in the commodity distribution network, including the losses that are caused by price reductions due to poor quality, the deterioration of products in storage, the additional rates on circulating assets resulting from the slow distribution of goods in the trade network, etc. Lower commodity distribution costs will increase profits and, consequently, the bonus fund for trade personnel.

All that we have said serves only to outline the general approach to the task of economic regulation of the enterprises' activities. And, of course, there must be detailed elaboration relative to the various types of enterprises: production, construction, trade, etc.

Objections may be raised about any of the above-mentioned methods. But then an alternative way of solving the problem should be suggested, for the need for organizing all-embracing economic influence is beyond any doubt.

Several articles have lately been published on the need to create a network of computing centers for national economic planning. An extensive use of computers in economic calculations should be supported in every possible way, for in this field we lag far behind the advanced capitalist countries. It would, however, be self-deception to consider that the problem of optimum planning and management can be solved with the aid of computers alone. Its solution must be based on correct economic criteria that will stimulate economic development along set lines.

We consider it expedient to commission a group of enterprising economists and economic executives to:

a) <u>work out a system for economically stimulating the opera-tion of enterprises</u> (production, building, and trade), <u>bearing in mind the transition from rigid, item-by-item restrictions to economic influence</u> (bonuses, taxes, fines, variable prices, interest rates on capital, etc.);

b) <u>determine the extension of the plant executives' powers;</u>

c) <u>determine flexible forms for the bonus system;</u>

d) <u>submit suggestions on the introduction of this system at some enterprises in the near future (by way of experiment).</u>

Pravda, September 1, 1964

What Is Useful for the Country Is Profitable for Everyone

V. Shkatov

The question of enhancing the importance and role of profit in the material incentives of the working people has been ripe for some time. At the 22nd Party Congress N. S. Khrushchev said: "We must increase the significance of profit and profitability. In the interests of better fulfillment of plans, the enterprise must be given greater opportunity to dispose of its profit, to use it more extensively to encourage better work by its personnel, and to expand production."

At present, bonuses to workers, engineers, technicians and office employees depend on the fulfillment of numerous indices. Moreover, bonuses to workers depend more on quantitative, physical indices than on the reduction of production costs and the raising of profitability. Bonuses to the engineering, technical and office staff, on the contrary, depend more on the fulfillment of plan assignments for reducing production costs. The personnel of enterprises operating at a planned loss, and many of them still exist, are more handicapped under these circumstances than personnel of profit-making enterprises. Meanwhile, it is no secret that planned unprofitability, especially in the extractive industries, by no means always depends on the work of the enterprise's personnel.

In his article "For Flexible Economic Management of Enterprises," Academician V. Trapeznikov has correctly raised the question of replacing the large number of standards, presently used as criteria for evaluating the economic efficiency of enterprises and for determining the size of bonuses, by a single criterion. This criterion can only be profit, for all the other

---

The author is the assistant head of a subdivision of the Price Bureau of the USSR State Planning Committee.

indices and results of the enterprise's operation intersect in it as in a focus. The growth of labor productivity and the reduction of production costs find their expression in higher profits.

Profits (or savings due to a reduction in production costs, in the case of enterprises operating with a planned loss) are presently the source of the enterprise fund; a part of this fund is used to improve the cultural and living standards of the personnel, while another part goes to improve production facilities. In our opinion, the first thing to be done is to change the procedure and scale of allocations to this fund. The fund cannot play an important role if its size is restricted to 5 1/2% of the wage fund. That part of profit which remains at the disposal of the enterprise's director for purposes of collective and individual incentive and for improvement of production facilities must be increased.

Once the role of profits in providing material incentives is enhanced, practice itself will show which indices need not be included in the plans drafted for enterprises.

In order to increase the significance of profits in the material incentives of working people, a series of measures have to be carried out. The point is that profit must properly reflect the operating efficiency of each enterprise.

This has exceptionally great importance for enterprises in the extractive industries. The point is that in the extractive industry, the productivity of labor, production costs and profitability substantially depend, first, on the quality of natural resources (mineral content in ore, the caloric value of mineral fuel, the sulphur and oil content of petroleum, the quantity and quality of timber per hectare of forest area, etc.), and, second, on the location of the natural resources relative to the earth's surface, transportation lines, and markets. Thus, for instance, the profitability of mining enterprises in the nonferrous metal industry varies, under the influence of these factors, approximately from −70% to +150%.

In order to remove the effect of natural and geographic factors upon profits in the extractive industry, it is necessary to introduce monetary payment for the development of natural wealth. The rates of payment should be differentiated depending on the quality and location of the natural resources. Such payments exist in the USSR's timber industry (stumpage). As of January 1, 1966, such payments are to be introduced in some iron-ore mining areas, and it is envisaged that they will be introduced for every ton of asbestos ore mined.

Many people believe that if paid exploitation of natural resources is introduced, a considerable rise of wholesale prices in heavy industry, with all the ensuing consequences, will be inevitable. But actually this may be avoided. In the petroleum, gas and asbestos industries, in hydroelectric power production, payment for natural resources can be introduced without changing the wholesale price level. In these industries, monetary payments can be introduced at the expense of the turnover tax on oil products, gas and electricity, and of the high profits obtained in the asbestos industry. The rates of payment for natural resources may (by analogy with "forest taxes" in the timber industry) correspondingly be termed "petroleum, gas, asbestos-ore, water-resource taxes."

In other extractive industries it would be expedient to carry out a differentiated evaluation (depending on quality and location) of the natural resources by zones and to equate the mean points on the evaluation scale to zero. Then the natural resources that are above the average will have a positive tax, while those that are below the average will have a negative tax. Let us call this system of evaluation "polar taxes."

The enterprise that exploits natural resources which have a positive tax has to pay a certain sum into the state budget. When an enterprise exploits negatively taxed natural resources, it draws a certain sum from the state budget. In this case, the enterprise's profit will reflect only the level of economic efficiency of its personnel and will not depend on the given quality and location of the natural resource. If the effect of natural and geographic factors upon profits is not eliminated by means of differential polar taxes, then the growth of the role of profits in materially encouraging the personnel of the extractive industries will worsen the conditions for the consistent implementation of the principle of equal pay for equal labor.

Moreover, the proposed system of monetary assessment of natural resources will become an important economic stimulus for a more rational utilization of natural resources. This is an important task which stems from the objectives set by the Party Program. We still have many shortcomings in our use of natural resources.

Intolerable wastefulness and unwarranted losses of mineral raw materials exist at ore-mining and metallurgical enterprises. Heavy damage is inflicted upon the national economy by the selective development of mineral deposits, in which only rich sites are exploited. Such a practice leads to irretrievable losses of useful minerals. The losses of coal, iron and manganese ores, nonferrous and rare metals, and many other useful minerals are immense. They constitute 15 to 25% of the stocks (without taking into account the losses sustained in dressing and the metallurgical remelting of ores).

The coefficient of oil recovery is particularly low. Specialists have calculated that an increase of this coefficient from 0.44 to 0.6 alone would make it possible to extract tens of millions of tons more oil annually without sinking any additional wells. There is still too little recovery of casing-head gas. Several billion cubic meters of this gas are burned in torches and discharged into the air every year.

Great quantities of pyrite with low sulphur and copper content are to be found in pit heaps, while in some capitalist countries sulphur and copper are extracted from pyrites of similar mineral content. At the quarries of the Mikhailovo Combine in Kursk Region, quartzites with a 42-43% iron content are used as ballast under railway track ties.

In recent years, measures have been taken to put the use of natural resources into proper order. Most union republics have passed laws on protecting nature. Some of them have established special agencies for nature protection. Those guilty of inflicting damage on natural wealth are held to account for such actions in courts or administrative bodies.

But all these measures suffer from being one-sided. They are not backed by economic measures, by collective stimuli to the enterprise personnel engaged in the exploitation of natural wealth.

The only way to solve this problem is to introduce differentiated polar taxes as payment for useful minerals. If the Mikhailovo Combine were to pay the state budget for quartzites, it would find it cheaper to bring ordinary ballast material from tens of kilometers away, and to allocate the money thus saved for the construction of a dressing factory.

The introduction of differentiated assessments and polar taxes on natural resources will create conditions for improving cost indices. Thus, for instance, payment for water resources and for the crop and other areas flooded in the course of hydroelectric power construction will permit the true production cost of hydroelectric power and its economic effectiveness to be evaluated in comparison with the power generated by thermal power stations.

Differentiated payment for land will obstruct the use of fertile land as construction sites and will make for more compact layouts of our enterprises and towns. Specialists believe that construction in our country is two or three times more wasteful than abroad.

Thus, Trapeznikov's proposal can be adopted by the extractive industry, and it will produce an important economic effect only if natural resources are assessed in monetary terms.

Pravda, September 7, 1964

The Plan and Methods of Economic Management

L. Leont'ev

In his article "For Flexible Economic Management of Enterprises" (Pravda, Aug. 17, 1964), Academician V. Trapeznikov has raised quite urgent problems. They have been discussed for a long time, but this has not made them less important or less urgent. On the contrary, life increasingly dictates their earliest possible solution.

Our economy has grown into a gigantic force and is making steady progress. It has inexhaustible possibilities and reserves at its disposal. The significance of initiative on the part of enterprises, their executives and personnel, in utilizing these possibilities is indisputable, as is the part played by material incentives in the development of initiative.

This is the essence of the problem: how can the planned centralized management of the national economy, the need for which is unquestionable, be combined with the greatest possible scope for the undoubtedly significant local initiative and independence at the enterprises? Can these two elements be combined? Does not an insoluble contradiction exist between them? It would seem that those who oppose a sweeping extension of the rights and independence of enterprises proceed from the unstated, and sometimes even unrecognized, conception that this problem cannot be solved.

Management has been aptly defined as primarily the art of choosing — choosing the shortest way to achieve a fixed target, the best way to utilize resources and the most expedient and most rational methods of activity. This involves not only choice,

---

The author is a Corresponding Member of the USSR Academy of Sciences.

but solutions as well.

When a state plan which has the force of law regulates, fixes, and predetermines all aspects of a plant's activity in the most minute detail, little room is left for choosing the means to carry out the plan assignments. An unwieldy system of plan and report indices deprives the enterprise of freedom of action, gives rise to an unchecked stream of paper work, and complicates and slows down the process of developing, examining, and approving plans. Does this system have any merit? Can a system of indices — even a most ramified one — embrace all the many-faceted activity of an enterprise? It is an illusion to think so.

It is generally recognized that the system of plan indices and, consequently, of criteria for estimating the enterprise's efficiency must be improved, and that it must become less bulky and more graphic. However, opinions differ as to how this should be achieved, what has to be done.

The functions of planned management in a socialist economy are extremely important and extensive. Planned management has to ensure the party's economic policy as regards high rates of expanded reproduction, the essential lines of economic growth (priority production of means of production combined with adequate rates of growth of consumer goods production, accelerated development of the most progressive and promising industries, rapid technical progress in the entire national economy, etc.), and proportional development of the whole national economy. But it would be naively utopian to think that these aims can be attained through administrative instructions to enterprises on the manufacture, distribution, and utilization of each and every nail. The targets of the planned management of the national economy are attained by a skillful combination of direct assignments to enterprises and the putting into operation of flexible economic methods, based on material incentive, for influencing their efficiency. This presupposes that the center of gravity be shifted from administration by injunction to economic methods of guiding enterprises.

For many years the party has been demanding that the industrial managers pay assiduous attention to the economics of production, to precise economic analysis and calculation, to the maximum utilization of the principle of material incentive. The Party Program lists, among many other powerful sources for accelerating the progress of the Soviet society to communism, such factors as thrift, the rational use of national resources, steady improvement of planned guidance and management methods, and development of the people's initiative.

Economic methods of planned management have the advantage of putting into action the power of material incentive: what is good for society is good for the enterprise. There is no such coinciding of interests with administrative management methods, when the enterprises quite often prove to be interested not in enlarging production, but in getting an "easier" plan.

The transition to the broad use of economic management methods requires a single criterion of evaluating their efficiency and, consequently, for stimulating materially the enterprise's executives and personnel. The general recognition that such a criterion as the fulfillment of the assigned gross output plan is useless has led to quests for other criteria. But none of the suggested alternatives (the normative cost of processing, for instance) reflects the entire scope of the enterprise's operation, and therefore they cannot be considered as exhaustive synthetic indices.

It is the enterprise's profit that constitutes such a criterion, if the enterprise fulfills its plan assignments for the output and assortment of goods and delivery terms, if it adheres to the existing legislation in the sphere of labor and its remuneration, financial activity, contracts, etc.

With adequate economic proportions in the national economy, which reflect the objective state of affairs, profit serves as the most generalizing criterion of the enterprise's entire activity, of its positive and negative aspects, of its successes and failures. Marx's remark that commodities are a natural leveler retains its significance to a certain extent under socialism too: profit, in monetary form, serves as the common denominator to which all the different enterprise expenditures and receipts are reduced.

Two objections are raised against taking profit as the major criterion of an enterprise's efficiency and as the basis for material incentive. It is alleged, first, that in the pursuit of profits the enterprises will produce goods that society does not need and, second, that profit could give a distorted picture since it largely depends on prices. There is something to these objections, but it seems to us that these dangers may be averted.

It is necessary to recognize that in a planned socialist economy, profit may serve as the major criterion of an enterprise's activities and as a basis for material incentive only given certain prerequisites and conditions. As previously stated, the necessary prerequisites include the fulfillment of plans for gross output, assortment, and delivery terms. A precise and economically substantiated evaluation of the total expenditures, on the one hand, and of the goods produced, on the other, is the most important condition. In other words, the use of profits as a criterion of an enterprise's efficiency presupposes the correct solution of the problem of payment for production assets and economically sound price formation.

The absence of payment for production assets distorts the real picture of expenditures for production, prevents these expenditures from being fully totalled, and encourages inefficient executives who demand ever increasing capital investments while the available equipment and premises are not fully utilized. The introduction of charges for production assets is the most important condition for the solution of the task that N. S. Khrushchev defined precisely in these words: "to achieve a maximum of output per unit of productive capacity with a minimum of expenditures."

One can readily see the consequences of free-of-charge assets from the example of a Moscow factory that has recently been discussed in Pravda.

For over ten years, the plant has been holding first place in competition, with excellent indices of plan overfulfillment, growth of labor productivity, and reduction of losses due to rejects. However, an economic analysis revealed considerable reserves for obtaining still better results: it turned out that in the main shop, whose fixed assets had grown 2.6 times in the last five years, the output of goods per ruble of fixed assets diminished from 135 rubles in 1960 to 116 rubles in 1962. If production assets had to be paid for, such dynamics would have affected the economic state of the enterprise immediately and would have attracted attention. The introduction of the principle that production assets must be paid for would serve as a powerful stimulus for the better utilization of these assets.

Recently, about two years ago, the suggestion of making profit — under certain conditions — the major criterion of an enterprise's efficiency and of introducing the practice of payment for production assets met with the allegedly "principled" objection that these were capitalist categories which were unacceptable for a socialist economy. Now such opinions are voiced quite rarely. It has become clear that these categories have an altogether different social content in our conditions and may be employed as important tools for the economic management of the national economy. Now there is every opportunity for a businesslike, practical discussion of these issues.

Another complicated problem, perhaps the most complicated one, arising in connection with the use of profit as the single criterion of an enterprise's efficiency is that of price formation. The price system is the point where all the threads of the planned management of the national economic entity come together and where the complex aggregate of intra- and inter-industry relations and links is coordinated. Moreover, economically substantiated prices are a most important condition, under which profit serves as a precise criterion of the work of each enterprise. The cost of production of goods depends on the prices of raw and other materials, on electric power and freight transport rates, while the enterprise's profit depends, in addition, on the prices at which its produce is sold.

The need for flexible and dynamic price formation is dictated by a number of tasks, including those of speeding up technical progress, guaranteeing correct proportions between demand and supply, stimulating the struggle for high quality of produce, finding substitutes for goods in short supply, and many others. But no matter how flexible the price policy may be, it must, of course, be based on definite principles.

The price must reflect the sum total of the socially necessary input of labor into the manufacture of the product. The determination of this sum total presumes that full consideration is given to all the expenditures of living and materialized labor — in other words, to the amount of labor and materials required for the manufacture of the given product. Opinions do not differ in this respect. But the kinds of goods differ as regards their capital-output ratios. The consideration given to the capital-output ratio in prices meets with the objection that such consideration contradicts the labor theory of value.

This objection is apparently based on a misunderstanding. Marx's theory of value teaches us that value is created by labor, but it does not maintain that price is always determined by value. Prices are directly determined by value in simple commodity production, i.e., in a primitive economy based on manual labor with the simplest implements, where the question of varying capital-output ratios has no relevance. Marx provided an analysis of the commodity as a product of capitalist production, and he showed that the price of such a commodity is determined not by value directly, but by its modified form — the price of production. An analysis of commodities as products of socialist production leaves no doubt that their prices must be fixed with due consideration to the products' capital-output ratio or, as economists often say, perhaps not very aptly, by the "price-of-production formula." The consideration of the capital-output ratio in price formation is an essential prerequisite for paid production assets. To recognize the need for paid production assets and to reject the need for considering the capital-output ratio in prices is, to say the least, tantamount to manifesting inconsistency.

These are some of the issues that merit discussion both on the basis of practical experience and in the light of the economic theory of communism, which the party develops and enriches by generalizing the activity of a people that is building the material and technical foundation of communist society.

Pravda, September 20, 1964

Once Again on the Plan, Profits and Bonuses

E. G. Liberman

In his article "For Flexible Economic Management of Enterprises" (<u>Pravda</u>, August 17, 1964), Academician V. Trapeznikov has raised the question of substantially increasing the efficiency of our production on the basis of a correct utilization of profits. This is a trend that many will welcome.

The changes introduced into the management of our national economy in accordance with the decisions of the November (1962) and subsequent plenary meetings of the Party Central Committee have created, in our opinion, particularly favorable conditions for <u>enhancing the role of economic methods in influencing production</u>.

For this purpose we may well use profit as a generalizing (though not the only) criterion of the economic efficiency of production. I am of the opinion that Academician Trapeznikov's suggestions ought to be refined in certain respects. We should not deal with profit in its absolute sum, because the larger the industrial enterprise, the greater its profits, all other conditions being equal. Therefore, to estimate efficiency, profit should be correlated with some basis that characterizes the capacity of the enterprise. <u>The value of enterprise production assets</u> could serve most adequately as such a basis. Then the estimate will proceed from the profitability of production, by which we mean the relationship of profit to the value of production assets. Evidently, we must have long-term standards for similar enterprises in each industry, which would determine the extent of material encouragement on the basis of the profitability level attained.

The 1962 discussion showed that the amount of material encouragement must be commensurate not only with the size of assets, but also with the sum total of living labor applied at each enterprise. This is an entirely feasible task. The bonus rate will depend on the level of profitability, but the rate proper may be expressed as a share of the wage fund or as rate per employee. With a profitability of 10%, for instance, the incentive fund would receive, from profits, 7% of the wage fund, while with a profitability of 20%, the incentive fund would receive 12% of the wage fund, and so on. These rates have to be carefully worked out. In this case the enterprises will not be interested in artificially increasing their wage funds, since that would lead to higher production costs and, consequently, lower profitability.

The chief and most serious objection to profitability as a criterion of production efficiency was that profitability often fails to coincide with actual production efficiency. Numerous statistical data were cited to prove the previously established fact that production costs, for instance, may grow while profits and profitability not only fail to diminish, but, on the contrary, increase. But that in no way proves that profit is not a suitable measure for estimating operating efficiency because the above contradiction is a result of defects in price formation.

Our prices are too often divorced from their natural basis—the socially necessary production costs, which are best measured by the average branch expenditures for one or another product of labor. For some reason or other, the principle of state plan prices was understood as the principle of permanent prices for all goods and for the same long term. This was the factor that created the diversity in the profitability of goods.

Now, however, more and more people are convinced that prices must combine stability with flexibility. Prices for individual groups of goods must be revised as the correlation changes between the average cost of production of a given branch and the operative prices, with due consideration for the new goods' novelty, quality, and effectiveness in operation or consumption.

Of course, it is not an easy or simple task to combine price stability with flexibility. But if it is approached from the position of rational and necessary decentralization, then it can be solved. We can make extensive use here of the principle of calculated intrabranch prices, which was employed in the prewar period and proved its value in many branches. Then the price remains stable for all consumers, but the producers, depending on their equipment or on natural-geological conditions, will be paid different prices so as to create equal conditions for profitable operation. Of course, we may also apply the principle of rent payments for mineral resources or for assets. All these problems are in urgent need of solution.

Only a flexible price policy and substantial reward for profitability can encourage enterprises to make the best possible use of their fixed and circulating assets, to steadily improve production methods, to put into production new types of goods, and to improve the quality of their produce. Meanwhile, methods based on numerous valuation indices, dictated from above by plan allotments, will hardly produce the desired results.

In our conditions, profitability is far from being the only criterion of efficiency. First of all, an enterprise's work has to be evaluated with due consideration for the quantity, assortment and quality of goods and their delivery dates. Contract deliveries, based on direct connections between the manufacturers and consumers, are the foundation of stable plans. This estimate of the suppliers' adherence to terms has to be supplemented by profitability. The failure to meet delivery terms should involve substantial fines. On the other hand, conscientious and enterprising suppliers should be given solid encouragement.

Such indices as the growth of output volume, higher finished output, greater output per ruble of fixed assets, lower production costs — all together and each individually — have their own significance in the planning and accounting process. But they all come together and cross in profitability which, for this reason, ought to be employed as the key criterion for estimating the efficiency of an enterprise's operation.

But this by no means signifies that all the diversity of branches and enterprises will be ignored. The general bonus fund will be formed on the basis of a single principle. But within the enterprises in different branches, and even at different plants, this fund may be used to give bonuses on the basis of criteria that are decisive in the given conditions. For instance, in the extracting and raw-materials branches, or even in billet shops, maximum encouragement should be given for increasing the volume of output. In instrument making, bonuses should be paid for raising the reliability of the instruments put out. In all these cases, the following rule must be maintained: encouragement should be based not on instructions that regulate in every detail

the process of estimating and rewarding, but on extension of the
enterprises' rights within the framework of a common fund and
general legislative standards. With excessive regulation of incentives, bonuses are often turned into a supplement, a guaranteed makeweight, to wages, and their stimulating role is reduced to naught. More encouragement should be given for individual achievements, and such bonuses should be paid currently and quickly.

In essence, all these principles are already being applied in
the operation of two sewing industry combines — the "Bolshe-
vichka" under the Moscow City Economic Council and the "Maiak" under the Volga-Viatka Economic Council. Similar experimental measures are also being prepared in the Ukraine. For
these enterprises, a plan is nothing but a projection of the
orders placed by stores for a given period of time. The enterprise drafts its own plan on the basis of these orders. Efficiency is estimated on the basis of profitability, while bonuses
are paid only out of profits, there being no regulation whatsoever as regards who is to draw bonuses and for what; this is
decided by the management and public organizations of the enterprise. All these measures pursue the aim of improving the
quality of goods and speeding up their sale.

It is useful to note that we are not talking about a revision of
indices, but about a reform of the enterprises' relations with
the national economy. That is why we find it quite strange when
people argue: what is better — the normative processing cost,
for instance, or profit? Profit must function as the general,
ultimate measure of efficiency. But the main point here is that
profit need not be fixed for enterprises in plans from higher-
level organs, just as other qualitative indices should not be
limited either. All these indices will necessarily be planned on
the scale of republics and economic regions, as well as of production branches. And the enterprises, on the basis of delivery
contracts and the rates of reward from profits, are quite capable of drafting their plans themselves. Moreover, the enterprises will find it profitable to work for maximum profitability
both in their plans and in actual practice.

Since profitability is the focus or the optimum point of inter-
section of all other qualitative indices, enterprises should seek
the expansion of production, the manufacture of new goods
(profitable prices!), new technological processes (lower production costs!), the best possible utilization of equipment as regards time and capacity, minimization of circulating stocks

(high profitability!), and improvement in the quality of the goods manufactured (guaranteed market and realization of profit). In other words, they will work for real production efficiency and not just for a formal attainment of indices.

Consequently, the use of the profit criterion is not merely the replacement of the other indices by the profit or profitability index. If profit is regulated by the plans fixed for enterprises, there is no guarantee that the latter will not, as they do now, often strive to understate their plans and thus slow down progress in production. That is why enterprises must have a long-term norm for bonuses that is not subjected to revision every time that the enterprises exceed it. The enterprises must always be assured that the personnel that have scored a certain success will enjoy a share of its results.

What is profitable for society, as represented by the state, must be profitable for every collective of the enterprise and every member of that collective! It is essential that every step forward should be of great benefit to society but, at the same time, that the given collective does not fail to get a certain share of this benefit.

I should like to say something about certain commentators in the USA and the Federal Republic of Germany. They have been interpreting the desire to make better use of profits in the USSR as a transition to a market economy and, even, to the system of "free enterprise"! Some snipes, as they sit in their bog, think that nothing in the world can be better and that all others must strive to live in the bog too. If these snipes had a little broader outlook, they would see that profit in the USSR can be a better measure of production efficiency than it is under capitalism. The point is that in our country profit cannot be increased through speculation, deliberate price rises, nonequivalent exchange with backward and colonial countries, and pressure upon the wage level. Our profit, provided that the prices reflect correctly the average branch costs of production, is nothing but the effect of the growth of social productivity, collected in money form. That is why we can pay bonuses for actual production efficiency, proceeding from the profitability criterion. But such incentive does not mean enrichment. In our society, profits cannot be turned into capital since no one — neither the director, nor the trade unions, nor any private individual — can use the bonus that he has received to acquire privately any means of production. So where does the "private enterprise," which those "snipes" have been heralding, come in?

Where is the "market economy" if we fully preserve centralized planning, which we want to improve and strengthen by releasing it from the details of petty control over the enterprises' activities and by integrating the Leninist principle of material incentive into the very process of planning?

What we are talking about is not capitalist "enterprise," but the strengthening of initiative on the part of our workers, engineers, and economic executives, and the encouragement of this initiative on the basis of the socialist law: "high labor remuneration for high achievements."

The numerous and frequently complicated problems that arise on the road to improving the methods of economic influence on production require, in our opinion, more serious measures than the establishment of a temporary commission of enterprising economists, as Academician V. A. Trapeznikov has suggested.

Apparently, what we need is not an interdepartmental office, but a superdepartmental and standing body at the highest level, which would direct all inspection measures and calculations and prepare, stage by stage, a system of legislative acts along a definite, firmly established line.

Pravda, November 13, 1964

The Chief Thing Is Economic Effectiveness

R. Belousov

A characteristic feature of Soviet economic progress today is the acute interest in questions of economic effectiveness. This is accompanied by a persistent search for better methods and means of national economic planning, as well as by a broad exchange of views on the subject.

Adherents of the conventional, though already antiquated forms of economic management, who lack convincing arguments against giving the enterprises more operative independence, propound the thesis that broader use of economic methods of influencing production collectives will subvert the foundations of centralized management of the national economy.

But their apprehensions are unfounded. First of all, it should be emphasized that the principle of centralized management of the socialist economy is an objective necessity which follows from public ownership. One can hardly find a single Soviet economist who would question the fact that only a unified state plan can express the interests of the whole people. This plan should establish, on a national scale, the targets for raising the living standard, securing the most important national economic proportions and progressive changes in the structure of social production, strengthening the defense capacity, furthering economic ties with the fraternal socialist countries, etc.

But after all these targets — which express our party's economic policy — have been elaborated, considered and endorsed, they must be communicated to the enterprises and construction sites. And there are two ways to do this.

One way is to cover in the state plan as many production ties as possible, to centralize all the available resources and manage the lower economic links by direct instructions, that is,

219

by administrative methods. This approach will require the plan to include a large number of physical and other quantitative indices. This type of planning took shape — for various reasons — during the period of the first five-year plans and in the war years. It still exists today, though it has been modified in various degrees.

The other way is to encourage the initiative of the production collectives in every possible way, to execute in deed, and not in words, planning "from below," communicating plan targets to the enterprises not only and not so much by directives, but through a system of orders, utilizing the direct connections existing between the enterprises, and exerting an active influence on them with the help of planned prices, state finances, and other economic levers that remain solely at the disposal of the central bodies.

At first glance it would seem that the first approach is simpler and more effective. However, an excessive increase in the number of plan indices not only will not strengthen centralized planning; it will actually undermine it. Important problems of socialist reproduction remain unsolved in the jumble of numerous specific questions.

For instance, many of the measures envisaged by the plans are not supported by economic efficiency estimates. The planners often fail to dovetail questions of material and technical supply and financial assignments with the production program. Not infrequently the plans for the current year are not received by the enterprises until February or March. And they have to be amended constantly. Thus, if one were to count all the changes that were introduced in recent years into the plan of the Moscow City Economic Council, it would emerge that, on the average, the various assignments were amended nearly every day. Unwieldy plans largely lose their organizing role.

The tenacious vestiges of subjectivism in planning are incompatible with the requirement set forth in the CPSU Program to the effect that the best results should be achieved at the least cost in the interests of society. The very wording of this law implies that economic results should be analytically compared with the outlays involved in achieving them. Can one say that all production and capital construction targets provided in the plan are based on such estimates? Unfortunately, the present tools of the plan — its indices — do not permit an objective comparison of economic results and the costs incurred by society. Therefore, these tools do not contain a sufficient economic substantiation of the economic measures that were carried out in keeping with the plan targets.

At times, owing to lack of coordination among the numerous indices, the production interests of the enterprise enter into contradiction with the interests of society. As a result, the national economy suffers considerable losses. For instance, certain metallurgical plants have no interest in collecting nitrogen, which they obtain in the production of oxygen for their own needs. So they literally cast it into the wind. And nearby chemical works liquefy air in order to obtain nitrogen, and let out the oxygen into the atmosphere.

The need for introducing economic incentives on a broad scale to stimulate production is gaining increasing recognition of late. Huge masses of people who are directly engaged in production have been drawn into the search for more rational ways of improving the economy. The use of economic stimuli will make it possible to achieve more effective and flexible direction of their efforts to utilize production assets in the best possible way, to consume raw and other materials economically, to introduce new equipment promptly, and to improve the quality of output.

In order to introduce economic stimuli on a broad scale, it will be necessary, above all, to:

determine accurately the indices of the effectiveness of the economic activity of the enterprise from the standpoint of the interests of society;

give the personnel of the enterprise an interest in reaching such indices; the personnel must be equipped with objective guidelines, with the help of which the work of the individual enterprise can be included most effectively and harmoniously in the process of socialist expanded reproduction;

curb significantly the employment by higher bodies of purely administrative methods of managing enterprises, and grant adequate independence to the enterprises in the selection of more effective ways of utilizing their production potential within the framework of the economic plan. Let us recall how Lenin put the matter: "There is nothing in common between democratic and socialist centralism, on the one hand, and the introduction of stereotyped patterns and uniformity from above, on the other hand. Unity in the basic, the fundamental and the essential will not be violated, but will be ensured by diversity in details, local peculiarities, and methods of approaching the problem. . . ."

The time is also ripe for arriving at a practical solution of
the problem of planning the economic effectiveness of capital
investments. We now calculate effectiveness only at the stage of
planning. However, when the problem reaches the stage of production, it recedes into the background or is overlooked altogether by the executives. One often hears talk about an enterprise not having reached its planned production capacity, but one
never hears about planned effectiveness not having been
achieved.

Many economists propose as the main index of the results of
the economic activity of the enterprise a norm which characterizes the relationship between the profit derived and the production assets. There is evidently no point in starting an ab-
stract dispute about the place this norm should occupy in the
system of other plan indices. Obviously, this norm should be
linked in one form or another with the index for assortment
[nomenklatura].

Only a large-scale economic experiment will provide an exhaustive answer to such questions. However, even now there are
grounds for asserting that, in comparison with the other indices,
the norm of profitability gives the most comprehensive and,
consequently, objective information concerning the results of
enterprise economic activity as compared with outlays.

Those who advocate an enhancement of the role of profit in
assessing the effectiveness of the results of enterprise activity
by no means regard it as a panacea for all kinds of difficulties.
The only advantage this index has over the other indices is that
it is the most general, summary index. This emerges most
clearly if profit is compared as an index with gross output and
commodity output, or with the standard cost of processing.
These indices suffer in greater or lesser degree from one-
sidedness and, hence, induce the enterprise in certain cases to
put out material-consuming products, and in others — labor-
consuming products.

An index of a more universal character is the production cost
target, for it takes account of the saving in current material
outlays and in wages. However, if the reduction of production
costs is advanced as the main criterion for assessing economic
activity, enterprises often save at the expense of quality. And
socialist society is not at all indifferent to the quality of the
output, to the reliability and durability of the machinery and
other articles used in the national economy. At present we are
confronted with a very unpleasant situation in which just as

many machine-building workers are engaged in repairing and adjusting equipment as in manufacturing new machines. In the metallurgical industry, one worker in every four is a repair worker. Numerous facts show that in recent years the operational qualities of machines have hardly improved, despite the very extensive technical possibilities in this area. One of the reasons for this is that in the pursuit of lower production costs, some plants use cheaper raw materials of poor quality, avoid technological operations requiring additional expenditure of labor to improve the quality of the product, etc.

The index of profits is more flexible in this respect. It fully reflects the economic effect derived both from lower production costs and from better product quality, inasmuch as produce with better consumer properties is generally assigned higher prices.

It should be borne in mind, however, that plan targets for securing higher profits and also for lowering production costs may be overfulfilled not only by improving the organization of the labor process and tapping other reserves, but also by making big capital investments. For instance, in the early years of the Seven-Year Plan period, huge outlays were made for the development of the iron ore industry. Taking advantage of this fact, certain executives of the iron ore administrations, seeking to lower their production costs and to increase profitability, put in requests for large quantities of equipment, although this was not always justified economically. As a result, the cost of ore extraction from very wet pits, as well as from pits with other unfavorable natural conditions, was lowered. But at what price? At the price of irrational capital investments. This example shows that the results of the work of the enterprise must not be assessed in a one-sided way. The profit and saving derived from lower costs of production should be compared with the magnitude of the production assets that were required to achieve this result.

The use of profit as an index of efficiency of the enterprise may raise serious objections in connection with the fact that the profitability of individual products varies greatly. The profitability of manufacturing individual items at one and the same machine-building or chemical plant often differs appreciably from the average level. But the fact that the size of the profit in the prices of certain types of items varies considerably does not indicate anything by itself. It seems that the level of profitability in prices will differ under any economic conditions. One of the shortcomings of the present prices is that differentiation of profit is not adequately substantiated from the economic viewpoint. For instance, it is difficult to explain why the level of profitability in the prices on pipes listed by the economic councils of the Russian Federation is nearly three times lower than the profitability in prices on rails. In such conditions the product which is most "advantageous" for the enterprise is not always the one for which the need is greatest in the national economy. But this is a defect of price formation and not of profit.

We hardly ever engage in the current revision of outdated prices and the regulation of profitability levels. The organizational forms of price formation are not adapted to the solution of such problems. The prices in force remain unchanged for years; they become outdated and enter into contradiction with the new conditions of production and marketing.

It is essential to arrange for systematic work on the regulation of prices, precluding any distortions in the price structure that are economically unjustified. It is important to make it most profitable for the enterprise to produce advanced types of products that will ensure the greatest effect for society as a whole. Only under these conditions will the economic levers work in accordance with the principle: what is good for the country is profitable for the production collective.

It should be emphasized, in conclusion, that improvement of economic management cannot be reduced to the questions of plan indices and price formation alone. Tackling questions of economic effectiveness also presupposes far-reaching changes in the present system of material incentives, as well as great changes in the system of material and technical supply, in the financial system, and in cost accounting. Of course, all this is very complicated and difficult. But once questions have matured, they must be solved.

Izvestia, December 4, 1964

The Independence of the Enterprise and Economic Stimuli

V. Belkin and I. Berman

Many of the economy's executives have high hopes for the statute on the enterprise and are waiting for this important document to appear.

In our view, the chief element in the statute should be the consistent realization of the universally recognized method of organizing and managing the enterprise — cost accounting, a method that is expected to ensure an organic combination of the interests of the state with those of the enterprise and individual workers.

Since the introduction of cost accounting, production has grown tremendously and the complexity of centralized planning and management has increased even more. True, our enterprises are now managed by experienced specialists, the cultural and technical level has increased, and the activity of the workers, engineers and office employees has intensified. It appeared that these factors required an expansion and strengthening of the economic independence of the enterprise. Of course, the economies of the regions continued to develop, but a cumbersome system of petty tutelage of the enterprise took firm root at the same time.

Yet the very concept of "cost accounting" implies that the enterprise itself should determine how to achieve the best results. But other agencies — the economic councils, branch committees and planning bodies — do this job for it. And it is established practice to criticize these bodies for unsatisfactory planning and management of the enterprises. It is high time to realize that guidance of the enterprises by means of administrative intervention in all the details is a very poor method that cannot yield good results.

There are people who believe that it will he possible to achieve completely centralized management of the economy with the help of computing techniques. This is an illusion.

The main line of work of the authors of this article is employment of electronic computers in the economy. We would very much like to see clever machines display their capabilities in this field too. But no matter how advanced electronic computers may become, a system under which everything — to the very last bolt — is planned and managed from a single center is an impossibility.

Moreover, electronic computers have hardly benefited economic planning and management as yet, mainly because current practices in economic management are not adapted to working out and, above all, executing optimum solutions.

Here is a case in point. It is now clear that the elaboration of optimum schemes for goods shipments yields a sizable saving. This is not a complicated task. It has been the subject of many articles and books, and not a few dissertations on the question have been defended. Yet there are hardly any shipments made under optimum schemes. Why not? Precisely because the transport agencies, contrary to common sense, are given plans expressed in ton-kilometers, and optimum schemes reduce the number of ton-kilometers. It is possible to set up all sorts of computing centers, to devise superb algorithms, but there will be no progress as long as the transport agencies are held accountable under the plan in ton-kilometers.

Since the question of computing techniques has been raised, it should be pointed out that the crux of the problem of introducing electronic computers involves economic preconditions — the need to reconstruct the system of planning and management and to create economic stimuli that will operate in the right direction. Only under these conditions will electronic computing techniques be able to produce (and will produce) a tremendous effect. Only then will a unified network of computing centers serve a useful purpose.

The methods of managing the economy should be economic in nature. We must devise a system in which the material interests of the enterprise as a whole and of each individual worker are served by the maximum utilization of reserves and potentialities, a system in which these interests will fully coincide with those of the state. What is good for society should be beneficial for each individual worker.

The existing practice in planning, calculating, and evaluating the results of the work of the enterprise satisfies this natural requirement only to a small extent. The fact that the interests of the enterprise and of society fail to coincide is an artificial condition which results from the work of the enterprise being appraised in accordance with the degree of fulfillment of numerous excessively minute and detailed plan assignments. When the work of the enterprise is evaluated according to the degree of fulfillment of the plan and planning is conducted on the basis of the level achieved, the enterprise is impelled to conceal its reserves and to ask for lower plan targets.

It follows that the work of the enterprise should be appraised not in accordance with numerous indices that are incomparable, disconnected, and often contradictory, but in accordance with one major index of a general character that will best reflect the result of the effort of the enterprise's personnel and management.

There is no need to invent such an index. It follows from the very essence of cost accounting, which consists in comparing and collating the results of production with the costs, in having the results not only fully cover the costs, but exceed them as much as possible. The difference between the results and cost is profit. The greater the profit, the better the work of the enterprise. Therefore, it is necessary to encourage and stimulate the obtaining of profit.

Nor is there any need to invent economic stimuli. They already exist in the form of bonuses. Therefore, bonuses and other incentives should be strictly dependent on the size of the profit derived.

The introduction of profit as the main index will also make it possible objectively to appraise the work of the apparatus of economic management. The size of the profit derived by the enterprises under the given management agency will help determine whether the personnel of that agency have worked well. Each executive will be appraised objectively and accurately.

Those who are against making profit the main index hypocritically allude to the capitalist system. But the evil of capitalism lies not in the drive for profit, but in its distribution, in the obtaining of income not in accordance with work performed but in accordance with the amount of capital owned. In our case, the effort to obtain profit will result in the creation of greater real benefits, in better satisfaction of the requirements of the working people.

Profit will be able to play its role as the main index only if the costs and results of production are correctly estimated. To-day the costs are not reckoned adequately; fixed assets are not fully reflected in them and circulating assets are not reflected at all. Society presents the fixed and circulating assets to the enterprise without charge. True, there are amortization charges on the fixed assets, although there is no such charge on cir-culating assets. But amortization is actually reimbursement of the value of the fixed assets, and the enterprise does not pay anything for the use of the fixed assets. When an enterprise takes money from a bank on credit, it not only returns the money, but also pays a certain amount of interest for the use of the money. Then why does not the enterprise pay anything for the use of the fixed and circulating assets? Since these assets are granted without charge, the enterprise does not try to make the best possible use of them.

It has been estimated, with the help of electronic computers, that one ruble invested in the expansion of fixed assets or material circulating assets now saves 15 kopecks of current outlays each year. It follows that if an enterprise fails to utilize its fixed assets or has superfluous circulating assets, it bur-dens the state with an annual loss of 15% of their value. It is es-sential economically to stimulate the most efficient use of fixed and circulating assets. And the best way to do this is to impose a special charge for the use of these assets.

The profitability index of an enterprise should be expressed in the size of the profit as related to the value of the fixed and circulating assets at its disposal.

Prices should also be worked out so that they incorporate the cost of production and a percentage of the value of the fixed and circulating assets. The existing prices do not comply with this principle, and their deviation from actual expenditures causes undesirable phenomena. Incorrect prices, for instance, constitute one of the main reasons why our economists over-looked the advantages inherent in the accelerated development of the chemical industry.

Estimates reveal that wholesale prices deviate substantially from actual costs, and in different degrees. For instance, the level of prices on fuel and power is 1.4 times below the prices on machine-building products, and the latter, in turn, are 1.6 times below those on items produced by the light and food in-dustries. Given such differences in price levels, the enterprise is not in a position to use them as levers to secure solutions that will be advantageous to society as a whole.

And it is not only the enterprises, but also the planning bodies that find it difficult to operate without economically substantiated prices. In order to determine whether one or another branch of industry is progressive, our economists often have to look across the ocean, and this should not be the only guide for them.

Our economists have long argued this question. More and more research workers and executives have been coming out in favor of prices that correspond to the actual costs of production. But those people in our state who are responsible for this important work pretend that this is not their concern at all. They approach the question of prices not from the standpoint of the interests of production, but from a primitively conceived financial point of view. One cannot help recalling "primitive mercantilists" in this connection. The unsatisfactory condition of prices has been noted by all the recent party congresses and Central Committee plenary meetings. It is time to ask why the party decisions have not been fulfilled, why public opinion is being ignored.

Extension of the independence of the enterprise and reinforcement of the role of economic stimuli do not mean that planning will be abolished or replaced. It goes without saying that the state will continue to effectuate a policy of planned capital investments.

The proposed measures should not be regarded as a single and final solution of all questions. The point at issue is only the creation of the foundation for further transformations, which should be introduced gradually after thorough preparation.

We believe that it would be advisable gradually to abolish control over the number of people to be employed and the wage fund. It is necessary to increase the norms of amortization, having included in them compensation for obsolescence. The enterprises should be given the right to amend the output plan with the consent of the customer. The plan should be based on orders from the customers.

It might be advisable, in establishing planned prices, to regard the prices as maximum prices which cannot be exceeded and to permit products to be sold at prices below those fixed. On the other hand, when an enterprise is producing a product of the best quality or a product with higher consumer properties than are stipulated by the standards, it should be given the right to fix the price with the agreement of the customer.

Substantial changes are also necessary in the field of construction. Construction organizations should not be financed on the basis of the volume of construction and assembly work actually performed, as is now the practice, but should be granted credits and payment only for completed projects that have been commissioned. And the rate of interest on the credit should be sufficiently high. If this is done the builders will not scatter funds over a large number of projects; they will be economically, that is, more effectively, encouraged to commission the projects as rapidly as possible.

The urgent need to introduce fundamental improvements into our methods of economic management is obvious. Open the latest statistical yearbook issued by the Central Statistical Administration and you will see that the effectiveness of capital investments in the national economy has dropped in recent years. One can easily imagine the losses incurred by the national economy as a result of this tendency. This is the price we pay for deferring the solution of questions that became urgent long ago.

It follows from the above that the question of the economic statute on the enterprise is by no means of local importance. To adopt the statute will mean to carry out a series of essential measures. This is completely possible and advisable, for these measures will bring our national economy incalculable benefits.

Finansy SSSR, 1964, No. 12

The Plan and Cost Accounting

G. Kosiachenko

The cognition and active use of economic laws are objective necessities of the planned socialist system of economy. The economic laws of socialism are effectuated by being made use of by society in a law-governed way through the planned guidance of the entire national economy. Under these conditions, the question of the extent to which our plans correctly reflect the objective economic laws is of decisive significance.

This is not just a theoretical question; it is a practical one as well. The quality of plans, the level of the scientific planning and management of production, and the correct use of economic levers by planning and managing bodies acquire paramount importance. The conscious use of objective economic laws signifies a tremendous enhancement of the role of the subjective factor in the economic development of socialist society. It is not a manifestation of subjectivism, or a voluntaristic application of any forms and methods of economic management. The organizational forms and methods of managing the socialist economy should be determined by the objective nature of the operation of economic laws; they should correspond to the nature of the objectively existing economic relations between enterprises, branches, and spheres of the national economy, and to their interdependence. At the same time, it is the improvement of the economic efficiency of socialist production that must serve as an objective criterion of the adequacy of the organizational and other forms and methods of economic management that are employed. All this points to the immense importance of improving the whole system of planned guidance of socialist enterprises.

Our press has noted the shortcomings of planning: the inadequate substantiation of some plan assignments; the absence of proper coordination of production and capital investment plans with the available material and financial resources;

insufficient consideration of the concrete conditions in which the individual enterprises function; frequent modification of the enterprises' plans, etc. However, diverse conclusions have been drawn from all this.

One of the chief issues that has evoked diverse interpretation is the problem of how best to combine centralized and decentralized planning, direct planning and economic regulation, and of what economic methods can promote greater efficiency of social production.

The complexity of the problem lies in finding the most correct practical combination of such seemingly opposite economic forms as the commodity-money form, i.e., indirect consideration of labor outlays, and the method of direct distribution of social labor on the basis of the operation of the law of planned development. Incorrect interpretation of these problems is most often due to a nonhistorical approach to their solution, to a nondialectic understanding of how an old form and a new content combine, and, sometimes, to a naturalistic interpretation of the law of value.

The mastering of commodity-money relations in the planned socialist economy does not involve merely reproducing the old mechanism of their operation or restricting it superficially; it means the establishment of a basically new mechanism, in which the system of economic levers connected with commodity-money relations functions as a tool of national economic planning. Value indices are not a mechanism of the spontaneous action of the law of value, but are wholly subordinated to the planning principle and serve as an instrument of the plan. Correct application of the economic laws of socialism and scientific technical economic substantiation of plan assignments play a decisive role in ensuring proportions in socialist reproduction. The improvement of commodity-money relations is part and parcel of the improvement of planning and of the whole mechanism of the planned management of the national economy. Under these circumstances we cannot speak of socialist production somehow being automatically regulated with the assistance of some individual value indices.

Some economists consider excessive centralization of planning to be the root of all evil and are beginning to look for ways of improving matters outside the unified national economic plan; they are trying to substantiate a special mechanism that functions automatically on the basis of profit alone, which is to be used for determining the relations among the individual enterprises. They consider that the main thing is to guard the enterprises against frequent interference on the part of higher-level bodies. In their opinion, the enterprises' concern for getting higher profits will serve as an adequate guarantee that such indices as average wages, production costs, capital investments, etc., will be correctly fixed and will always conform to national economic proportions.

We find this kind of setting-off of centralized directive planning against economic regulation with the aid of value levers in Academician V. Trapeznikov's article "For Flexible Economic Management of Enterprises" [Za gibkoe ekonomicheskoe upravlenie predpriiatiiami]. (1) Some of the suggestions put forward in the article, and aimed at strengthening cost accounting, improving the use of value levers, etc., are correct in themselves. But when the author suggests "giving up the outdated forms of guiding the economy that are based on directive norms," and converting economic regulation into the chief method of economic management, then such contraposing of straight directive planning to economic regulation can only weaken the planning principle in the development of our economy. In the conditions of a socialist economy, value levers can be correctly used only when combined with straight directive planning.

The problem is to find better forms of relations among enterprises, planning bodies, and credit and financial agencies, relations that will ensure a reinforcement of centralized management of the economy and the development of initiative among the local bodies and enterprises, as well as enhancement of the material and moral stimuli acting upon their personnel for the better utilization of material and financial resources.

The socialist nationalization of the means of production on the scale of the entire society makes the direct planned links between socialist enterprises, based on the law of planned proportional development, the main and determining form of relations among these enterprises. The national economy cannot exist as a single planned economy without these direct relations. In these conditions, any other system of relations among enterprises, which could itself — without a single state plan — automatically establish ties among them and fix national economic proportions, is unthinkable. Direct planned relations

among enterprises, i.e., direct planning, are not effected apart
from commodity-money relations; they do not represent the
direct accounting and distribution of social labor in its natural
form in contradistinction to value calculation. The essence of
the matter is that direct planned relations among enterprises,
as expressed in definite qualitative and quantitative terms, are
effected in the form of commodity-money relations. They are
not two equivalent forms of relations existing side by side.
They cannot be regarded as being equal, but neither can they
be put in opposition to each other. At the socialist stage of
development, direct planned relations among enterprises cannot be effected in any other way than by the use of commodity-
money relations; the planning principle itself cannot be effectuated outside the dynamics of the commodity-money
form.

I. Malyshev questions the correctness of the thesis that
direct planned relations among enterprises is the principal
determining form of relations and that the commodity-money
form is a secondary and subordinate one. He ascribes to his
opponents the view that production may be planned without the
commodity-money form, and only with the aid of physical indices, that a plan can be drafted without "rubles," with value
relations coming into action only later. (2) This strange interpretation has been invented by the author himself. It goes without saying that a plan is drafted both in physical indices and in
"rubles." And the whole dispute stems from the author's rather
strange conception of the nature of the relationship between
direct social labor and the commodity-money form.

As we know, the chief characteristic of socialist labor is
that it functions directly as social labor, while the commodity-
money form plays a subordinated role and is used in a planned
way in the interests of communist construction. Labor that is
distributed in a planned way acquires from the very beginning
the character of social labor directly, i.e., in the very production process, prior to the realization of the product of this
labor. Meanwhile, in the sphere of circulation, each product
undergoes additional social checking. The conversion of finished
goods into money, the marketing of the goods, is at the same
time a check-up on how adequately social labor is distributed,
on the extent to which the outlays of living and past labor correspond to social standards, on how well all the elements of the

national economic plan are balanced and meet the requirements
of the planned development of the economy. The appearance of
reciprocal indebtedness among economic bodies, delays in re-
paying bank loans, the existence of above-plan stocks of un-
marketed finished goods, cases in which enterprises fail to
redeem the equipment they have ordered because they are short
of funds, etc. — all these things confirm the need for such additional checking in the sphere of circulation. It serves as a
supplementary form of economic control and, at the same time,
as a form of checking the correctness of the plan itself and of
estimating, with the aid of value levers, the degree of its
adequacy and fulfillment.

Failure to understand this peculiarity of commodity-money
relations in a socialist economy, as distinct from private
commodity production, leads to the negation of any kind of historical continuity between our money and the money of preceding socio-economic structures. "Commodity-money relations under socialism," Malyshev writes, "have nothing in
common with commodity-money relations of the past, with the
exception of their outward, material form." The only thing we
have inherited — he explains — is that paper money is made of
paper and coins are made of metal, just as scales, weights and
cash registers have come down to us from capitalist trade. (3)
On the basis of this "methodology" the author actually reduces
the role of money to a purely technical tool for taking account
of labor, while such conceptions as universal equivalent, a
form of indirectly accounting for the outlays of social labor,
etc., are omitted from the characterization of money. Meanwhile, in speaking of the adaptation of the commodity-money
form to new economic conditions, it must be remembered that
money remains a form of indirectly accounting for the outlays
of social labor and differs basically from the taking stock of
labor directly in physical units, which will occur in the highest
stage of communism. The fact that the commodity-money form
is used by society for directly expressing social labor involves
a unique contradiction that is connected with the imperfect
nature of the commodity-money form as an economic tool for
planning the national economy. This is expressed in a certain
independence of movement of money resources in relation to
the movement of material resources, which gives rise to the
possibility of such partial disproportions as, for instance, a

discrepancy between the volume of capital investments and the provision of construction projects with materials and equipment, or between the growth of the population's incomes and available commodity resources, including the services offered to the population for a fee. This contradiction manifests itself, when production volume and labor productivity are being determined through value indices, in the appearance of so-called "profitable" and "unprofitable" goods, in an inaccurate and sometimes even distorted expression of the actual expenditures and results of the enterprises' activities in terms of such generalizing indices as production costs, profits, etc. That is why enhancement of the role of value indices must be combined with improvement of their utilization and further strengthening of the planning principle in the national economy, with improvement of centralized planned management and expansion of the rights vested in the local planning and economic bodies, with the consolidation of cost accounting at the enterprises.

Cost accounting presupposes a combining of centralized planned management of each enterprise and of the entire national economy with some economic independence of enterprises within the framework of a single national economic plan. The limits of economic independence must be established so as to give the enterprise an opportunity to manifest initiative and interest in making the most rational use of its material, labor and financial resources, and, at the same time, to promote planned relations between the enterprises and the entire national economy. This can be attained only if state assignments play the decisive guiding role in all the activities of the enterprise, if the interests of the whole society are given priority over those of the enterprise and, at the same time, the enterprise retains certain rights and initiative.

The national economic plan is not just the sum total of the plans of enterprises and economic councils, for these plans alone are not enough to guarantee the most efficient pattern of all social production and distribution, not to speak of the fact that these plans are not balanced in relation to each other. The determining element is the centrally elaborated national plan which, relying upon the plans of enterprises, economic councils and departments, corrects them and establishes the parameters of the future development of social production in accordance with the concrete historic tasks confronting the socialist society in the given plan period.

Some economists support their views with the following

argument: that which is profitable for society as a whole must also be profitable for all its members. But they draw different conclusions from this statement: some hold that the interests of self-supporting enterprises are absolutely identical with the interests of the state, as if this community of interests is assured automatically. Meanwhile, there is not and cannot be any such identity under the conditions of socialism. By acknowledging a certain separateness of self-supporting enterprises and, thus, the need to vest them with certain rights, we thereby recognize a certain separateness and specificity of these enterprises' interests. The interests of localities, of self-supporting enterprises, of enterprise collectives, and of individual employees coincide with the interests of society as far as the main, decisive matters are concerned. But this does not exclude their having some specific interests of their own. It must always be borne in mind that these specific interests play a positive role in socialist society: they are bound up with material incentive.

At the same time, these specific interests and their separateness may, in the conditions of the enterprises' independent economic activities, give rise to the possibility of their clashing with the interests of society as a whole. This occurs when the enterprise begins to be guided in its activities only by its own narrow interests to the detriment of those of the state, when local interests start gaining priority over the interests of the whole society. The effect of material incentives can coincide with the interests of all society only if these incentives and the activity of the enterprises are coordinated with the interests of society. And this cannot be achieved by letting things drift, or automatically as, for instance, by merely establishing some long-term scale of profit distribution; it presupposes the existence of a definite system of day-to-day planned economic guidance and control on the part of the state. Such problems as, for example, the establishment of wage and salary rates, the ensuring of a correct relationship between labor productivity and wages, the determination of the volume and sources of capital investment financing at enterprises, etc., must be solved on the basis of the interests of the whole national economy, and not in an isolated way, proceeding from the interests of the individual enterprises alone.

A self-supporting enterprise must have a certain profitability, but not a profitability that is attained by any means, even at the expense of society as a whole. The drive for profitability must be in accord with the interests of the entire state. It is this that expresses the connection of cost accounting not only with the law of value, but also with the law of planned development and other economic laws of socialism. That is why the limits of the enterprises' separateness and economic independence must be determined by the interests of the state as a whole.

When enterprises were being converted to operation on a self-supporting basis, Lenin particularly stressed the necessity of fully protecting the state's interests: "If we, having established trusts and enterprises on a self-supporting basis, are not able to fully guarantee our interests in a businesslike, merchantlike manner, then we shall be nothing but utter fools." (4)

The interests of the state, i.e., of the entire national economy, constitute the chief criterion in solving the concrete questions involved in the relations between state agencies and self-supporting enterprises. It is on this basis alone that the concrete forms of relations between enterprises and the higher-level bodies can be improved and that cost accounting and material incentive at enterprises can be promoted. Thus, ignoring a certain separation of the interests of self-supporting enterprises and their personnel from those of society as a whole is bound to lead to an underestimation of the system of centralized planned management, which alone can guarantee a proper coordination of these interests.

The socialist state, through its planning and economic bodies, guides the national economy and every enterprise with the help of a system of economic levers and plan indices which take due account of the specific features of each economic branch. These indices cannot be replaced by any one value index. Prices in our country are planned, and they cannot automatically react to changing production conditions or to partial disproportions in social production. The reaction to such phenomena is effected by means of direct planned management and corresponding measures, including, when necessary, the planned alteration of prices. That is why any sort of automatic regulation of the enterprises' activities through price and profit fluctuations is out of the question under the conditions of a socialist economy, for that would mean a transition to spontaneous price formation. Automatic regulation of the economy by means of profit shifts cannot substitute for direct planned guidance. The classical mechanism of the spontaneous regulation of social production through price fluctuation has already been greatly undermined in the conditions of monopolistic capitalism. But monopolistic capitalism cannot transcend the limits of the operation of the law of value and is incapable of producing a different mechanism for regulating social production, one that could guarantee the planned development of the entire national economy. Competition, though it is the antipode of monopoly, is not eliminated by the latter, and continues to exist "over it and alongside of it, thereby giving rise to a number of very acute antagonisms, frictions and conflicts." (5)

The fact that value categories no longer function as the spontaneous manifestation of the law of value, but are used in a planned way and serve as economic levers of the national economic plan, is one of the advantages of the planned socialist economy based on the dominance of public ownership of the means of production.

That is why profit, despite its great importance for evaluating the activity of an enterprise and the need to enhance its stimulating role, cannot serve as a sole and universal index which determines the direction of an enterprise's activity. This direction is determined by the sum total of conditions created for the given enterprise by the entire planned socialist economic system. This includes coordination of the development of the given enterprise with that of other enterprises in the given branch and other enterprises in the overall system of social division of labor, centralized capital investments, the supply of materials and equipment, etc. The accelerated development of any new progressive branch — for instance, the chemical industry — relies upon the entire system of economic levers of national economic planning which the state uses for the corresponding distribution of material, labor and financial resources. Naturally, historic tasks of so great a scale cannot be solved without enforcing the role of centralized resources and, consequently, of centralized guidance. Thus, we must increase the role played by value levers, particularly by profits, but this by no means requires that profit be transformed into the sole, universal index and that directive planning be replaced by some special mechanism of automatic regulation.

***

What ways are there for improving the quality of the plans that are fixed for enterprises? The plan, with its system of basic indices, and the degree to which it is carried out constitute the initial and determining criterion for estimating the activity of an enterprise. Consequently, the bonus system must be patterned in close connection with the fulfillment of plan assignments, with due account of the larger assignments accepted by the enterprise when the plan is drafted. But this presupposes improvement of the plans fixed for enterprises. No improvement of the bonus system alone and no extension of the enterprises' rights as self-supporting units will produce the desired results if the quality of plans is not radically improved. The inadequate dovetailing of the individual elements of the plan, interruptions in supplies, and partial disproportions or discrepancies not only in the plans of the enterprises, but also in those of economic councils and departments, lead to frequent changes in plans and undermine the personnel's confidence in the plan as a directive. An effort must be made to have no disproportions whatsoever in the plan proper — from the overall plan of the national economy to the plan for every enterprise — and to improve the method of the balance dovetailing of plan assignments. A well drafted plan — internally balanced, fully mobilizing the reserves of production and reflecting the rise of the technical level of production and the quality of output — is the most important prerequisite for improving the efficiency of production.

An improvement of branch planning and of balance dovetailing between branches and enterprises, which will guarantee the proportional development of the national economy, and a related improvement in the use of fixed assets are of great importance in present conditions. It must be said that branch and interbranch planning has deteriorated in the last few years. Meanwhile, the branch structure of social production is one of the most important proportions of the entire national economy. The creation of large branch production associations not only within the framework of individual economic areas, but also in a number of branches within the framework of the entire national economy, as well as the improvement of production specialization, are important conditions for improving branch planning.

The establishment of large branch production associations would make it possible to improve rate-setting within branches—

the elaboration and introduction at enterprises of advanced norms for the expenditure of raw and other materials, the use of production capacities, and the expenditure of labor per unit of output — and would make for better control of adherence to these norms. And this would make it possible to raise the level of the technical-economic substantiation of the plan. Such associations — patterned along the branch or integration principle — would be an important link in combining centralized and differentiated planning and management of the economy; they would considerably strengthen branch planning and interbranch relations, create favorable conditions for the development of specialization and economically substantiated cooperation among enterprises, and ensure more stable production relations between suppliers and consumers. Finally, the creation of such associations would greatly facilitate the pooling of information and experience by related enterprises. At the present time, many related enterprises, even large ones, that are subordinated to different economic councils do not maintain regular contact; the executives of plants do not visit related enterprises of other economic councils in order to draw upon their progressive practices. A regular flow of information would make it possible to conduct analytical work on the scale of the entire branch and to apply more extensively the method of comparison and correlation both in plan drafting and in estimating the efficiency of enterprises.

Issues involved in the system and method of supplying materials and equipment, particularly questions of funding, the development of direct contracts between enterprises, etc., have been subjected to lively discussion of late. Inadequate coordination of output assignments and supplies of materials and equipment is one of the reasons for the frequent changes in plans at enterprises. In this connection, some economists propose to replace the funding system for supplying materials and equipment by a system based on trade through shops or in the form of direct contracts among enterprises for sale and purchase, including the purchase of items that come under the heading of fixed assets.

It must be pointed out that funding in no way excludes contract relations. The system of contracts should be developed and acquire the nature of permanent relations among enterprises. We cannot abandon the system of funding altogether and pass over to conventional trade. It is necessary to narrow down gradually the range of goods subject to funding and to shift, as regards materials that are in adequate supply (above all, those that are in mass production, such as all-purpose tools, spare parts, etc.), to marketing through trade and to deliveries under bilateral agreements. We must also improve the elaboration of materials balances, which are an important tool for securing, in plans, proportionality as regards materials, as well as for effectuating production relations among branches and enterprises. In general, the application of the balance method in production and financial planning, as well as the elaboration of synthetic balances, should be extended and improved. This method permits us to bring out the specific discrepancies between branches and spheres of the national economy, particularly between material and financial resources, while the plan is being drafted and to provide in the plan itself for measures that are essential to secure optimum proportions.

In speaking of the organizational forms and methods for the distribution of industrial goods, one cannot ignore the objective peculiarities inherent in the methods for distributing means of production and objects of consumption. Productive consumption is known to represent another aspect of production; the volume and pattern of this consumption are determined mainly by the production plan proper. That is why the methods and forms of distributing particular products of heavy industry depend to a great extent on the nature of these products, on their quantity in relation to the demand for them, on what consumers they are intended for, etc. A substantial number of these products, by their material composition, are designed for a definite consumer only and are distributed directly by means of contracts. However, the distribution of materials and articles in mass production, which are used in many production branches, must be more flexible and should depend, first of all, on whether, and to what extent, these goods are in short supply or there are enough of them to satisfy fully the demand of all consumers.

The distribution of consumer goods is quite another matter. In this case the product leaves the sphere of production and enters the sphere of personal consumption. (6) Here we deal with many millions of individual consumers with diverse and changing tastes, with regional and national peculiarities, and with constant change in the pattern of demand in accordance with the extent to which it is satisfied.

Along with quality, such factors as fashion and style are
becoming increasingly important with respect to many manufactured consumer goods. Commodities that are not in demand
among the population pile up at warehouses and stores. There
is now an above-norm stock of goods worth 2.5 billion rubles
in the trade network, most of which consists of clothes, woolen
and silk textiles, and haberdashery. In this case the marketing
problem is not one of an overproduction of goods that does not
correspond to a limited demand caused by a narrow consumption base, as is the case in the capitalist economy (no such
contradiction is possible under socialism); the problem here
stems from the manufacture of goods other than those that are
in demand. The result may be the creation of some unsatisfied
effective demand and, at the same time, the growth of stocks
of unsold goods. In these conditions, the market mechanism
should be resorted to on a larger scale: the methods and forms
of conveying the goods to the consumer must be more flexible.
It is also important to intensify the influence of the changing
demand upon production.

A well organized study of the population's demands by
trading organizations, which should be the ones to determine the
assortment of goods an enterprise must produce, is a decisive
factor in this matter.

It is the shop and the corresponding trade organization
that are primarily responsible for studying the demand and for
placing an order with the enterprise. But the enterprise must
also bear responsibility for the output of unmarketable goods
which do not conform to the order. This requires the establishment of more flexible relations between the enterprise and the
trade establishment. The enterprises of such branches as the
textile, clothing, and shoe industries should be granted more
extensive rights in planning the volume and assortment of their
output. The enterprise should itself, in agreement with the trading organizations, establish and, when necessary, change the
assortment, styles and models of the goods they manufacture
in accordance with the changing demand. It would evidently be
expedient to reduce considerably the number of plan indices
fixed by higher-level organizations for these branches, while
fulfillment of the plan for the marketing of goods produced and
of the profit plan should serve as the chief indices in estimating
the efficiency of the enterprise and in granting bonuses. A
thorough study should be made of the experiment now being
conducted along the lines of establishing direct relations be-

tween shops and the "Bol'shevichka" amalgamation of clothing
factories under the Moscow City Economic Council and the
"Maiak" amalgamation under the Volga-Viatka Economic Council, on the basis of orders placed by these shops. This experiment ought to be extended to include a larger number of enterprises, for this alone will make it possible to solve the problems of improving the planning of the branch as a whole, and,
in particular, the forms of production relations between the
branches of light industry and the national economy, the
methods of supplying materials and equipment to these branches
with due consideration of the actual possibilities, etc. It is
clear that the production plans of light-industry branches will
not just sum up the individual plans of the various enterprises,
drafted on the basis of the orders placed by shops.

***

Enhancement of the quality of plans presupposes a more
flexible use of economic levers in guiding the economy, and
the further development of the value and physical indices of
the plan.

The adjustment of wholesale prices is of great importance
in perfecting value indices. A price serves as the general form
of expressing the socially necessary expenditures of labor
under commodity-money relations. The unsatisfactory state of
prices and the diversity in the wholesale prices on some goods
have an adverse effect not only on such value indices as gross
and commodity output, production costs, labor productivity in
terms of money, etc., but also on many physical indices, such
as the assortment of goods, output quality, and others.

The introduction of new wholesale prices on January 1,
1966, will go a long way toward removing the diversity in
profitability and thereby weakening the tendency of some enterprises to exceed their plans as regards goods that are more
"profitable" to the enterprise but have secondary importance
for the national economy. It will also give the enterprises a
greater interest in putting out new goods and improving their
quality. This will enhance the role of profit somewhat. (7)

However, in order to have profit really function as one of
the principal generalizing indices for evaluating the work of an
enterprise and granting bonuses to its staff, it must occupy a
corresponding place in the system of plan indices. An enterprise obtains its accumulations from the sale of its goods. It
may happen that the output plan is fulfilled, but the goods accu-

mulate at the warehouse. That is why the efficiency of an enterprise should be assessed not by its gross or commodity output, but by its fulfillment of the plan for the marketing of finished goods and its accumulations. This applies both to the enterprises of light industry and to those of heavy industry. Implementation of this principle would promote improvement of the quality and assortment of output and would increase the enterprises' responsibility when they conclude business contracts. It is essential to increase the role of profit, first of all, as a most important plan index and criterion in assessing the work of an enterprise and, second, as the source from which the enterprise's bonus fund is formed, and also as a source for the further improvement and expansion of output at the given enterprise. Profit, as the most important element of cost accounting, must stimulate the fulfillment of plan assignments, the improvement of output quality, and the production of new goods.

Profit is determined on the basis of the goods marketed, and not on the basis of those produced. This is the factor that distinguishes it from, and makes it superior to, production costs as a criterion of an enterprise's efficiency. At the same time, profit, as distinguished from production costs, is affected not only by the relationship of prices on interchangeable raw and other materials, etc., but also by the relationship of prices on the different types of goods produced by the enterprise. This is due to the appearance of "profitable" and "unprofitable" goods, which affect the profit level. Of course, there are some contradictions here, just as with many other economic phenomena, that are connected with the existence of commodity-money relations. For instance, from the point of view of marketing, it is the goods that are in demand that should be manufactured, but these are sometimes less profitable than goods that are not in demand. This phenomenon may be observed in cases where we have obsolete prices. That is why the enhancement of the stimulating role of profit presupposes more flexible prices, that there not be a considerable discrepancy in the profitability levels of the individual goods, that prices be subjected to timely revisions in accordance with the changing conditions in the manufacture and marketing of the particular goods, and that the improved quality of goods be reflected in the price or in an economically substantiated addition to the price. It may be that this will require some modification of the existing system of current price regulation and periodic price adjustments, with the necessary degree of centralization preserved. But with all that, and regarding profit as a most important index for estimating the efficiency of an enterprise and for awarding bonuses to its personnel, there still remains the need to determine the degree to which higher profits are due to the efforts of the personnel of the given enterprise and the degree to which they are due to price and other factors. The profitability level, under conditions of uniform prices, also depends on the amount of equipment per unit of labor and the pattern of fixed assets at particular enterprises within the branch, as well as on other factors. But these questions merit special treatment.

The correct differentiation of plan indices is an important factor in improving the planning of the national economy and of each enterprise. These indices cannot be identical both for the state plan of the USSR and for the plans of the republics, economic councils, and enterprises. Not only the assortment and volume of the goods put out, but also all the other indices, including quality indices, should reflect the specific features of the branch, economic area and, in some cases, the enterprise.

In addition, it is essential to maintain a unity of quantitative indices expressed in physical and monetary terms. The measures of output employed in our plans must reflect more accurately the differing economic efficiency and the different qualitative properties of identical or interchangeable goods. Indices that ignore the use value of a product, its quality, cannot adequately reflect the socially necessary outlays and are capable of violating national economic interests.

Improvement of the physical indices of a plan must, as a rule, be accompanied by a change in the corresponding value indices. The consequence of a violation of this requirement may be indicated from the example of the paper industry, which introduced an index of paper output both in tons and in square meters, but left the calculation of paper production costs unchanged — in tons. The result was that the enterprises which succeeded in cutting the weight of a square meter of newspaper and printing stock and, thus, in considerably increasing its output, were deprived of bonuses because the overfulfilling of the production plan for area was accompanied by higher production costs per ton of paper.

Improvement of the quality of the plans fixed for the enterprises requires, first and foremost, a concrete approach to each enterprise. At present, the specific features of each enterprise are definitely not given sufficient consideration when its plan assignments are being fixed. No matter how perfect the branch norms and other indices that characterize the given branch as a whole (and, by the way, we have a long way to go in this area), their application to each enterprise presumes that the objective peculiarities affecting the production assignment level are taken into consideration. A plan can be realistic and, at the same time, capable of mobilizing reserves only if the specific features of every enterprise and of its operation in the planned period are adequately taken into account. The application of any sort of mechanical allocation of the plan assignment on the basis of average calculations or of the growth rate attained in the preceding plan period is a distortion of planning and testifies only to the poor work of planning bodies.

In determining the volume of output, for instance, one must concretely take into account the features of the given enterprise's operation: whether it is starting the manufacture of a new product or passing over to its serial output; the degree to which the technical norms have been mastered; how adequately the enterprise is provided with raw materials in the coming year and what its output should be; the extent to which its capacity is utilized and, in particular, how long it takes the enterprise to attain its rated capacity, etc. Only a plan whose sections are all substantiated by technical-economic calculations is capable of correctly reflecting the real possibilities and assessing the successes scored by the enterprise's personnel.

The personnel's interest in getting higher plan assignments is of importance in improving the level of planning at the enterprise. No matter how well the plan is drafted, it cannot disclose and take into account all the latent reserves of the economy. The personnel of the enterprise and its engineering and technical staff are intimately aware of their production reserves. But this implies an appropriate extension of the rights of higher-level economic bodies in the sphere of estimating the activity of enterprises and fixing the conditions and size of bonuses within generally established limits, as well as an extension of the rights of enterprise directors in the area of granting bonuses to individual staff members, depending on the role and participation of individual employees or groups of them in the attainment of production successes in particular sectors. The decision adopted by the State Committee on Questions of Labor and Wages and the All-Union Central Council of Trade Unions to introduce in 1964, as an experiment, new Regulations on Bonuses for Engineering and Technical Personnel and Office Employees at some enterprises of a number of economic councils was a step forward in this area. The Regulations provide for the differentiation of incentive criteria by branches and groups of enterprises.

However, concrete consideration of the unique features of individual enterprises in establishing plans does not mean that backward methods and levels of operation, which have taken root at some enterprises, can be tolerated. Assessment of the efficiency level of an enterprise and of the opportunities for its further improvement should be based on a comparison with the technical-economic norms of the branch or relevant group of enterprises, with due account of the technical features of the operating fixed assets and comparison with the achievements of other enterprises in the given branch. The norms fixed for enterprises with identical production conditions must be calculated for an identical level of labor productivity, regardless of the administrative subordination and territorial location of the enterprises.

Footnotes

1) See Pravda, August 17, 1964.

2) See I. S. Malyshev, Ekonomicheskaia nauka i khoziaistvennaia praktika, "Ekonomika" Publishing House, 1964, pp. 39-41.

3) Malyshev, op. cit., pp. 16-17.

4) V. I. Lenin, Soch., 4th ed., Vol. 35, p. 468.

5) Lenin, Soch., 5th ed., Vol. 27, p. 386.

6) This applies to all goods supplied to the population, including materials and goods for household and everyday uses.

7) We do not deal in this article with the debatable issue of the principles of price formation.

Ekonomicheskaia gazeta, January 6, 1965

The Firm and the Customer

V. Sokolov, M. Nazarov and N. Kozlov

Raising the living standard and the cultural level of the population is the prime concern of our party and government. The output of fabrics, footwear, clothing, cultural items, everyday necessities, and household goods is growing from year to year in the Soviet Union.

"The output of consumer goods," we read in the CPSU Program, "must meet the growing demand in full, and must conform to its changes. The timely output of goods in accordance with the varied demand of the population, with consideration for local, national and climatic conditions, is an imperative requirement for all the consumer goods industries."

Both industry and trade must be guided by these instructions. In order to take the changes in demand into account, demand must be studied systematically, in a planned manner, and on an economically competent basis. Unfortunately, both light industry and trade are still frequently doing this important work unsatisfactorily. And the result is an increase in above-norm stocks of goods.

One of the causes of overstocking is the imperfect way in which output is planned and distributed.

The system of centralized planning and distribution of commodities was correct, of course, when they were in extremely short supply. But times have changed. And every customer naturally wants goods that are not only made of high quality materials. A coat or a suit must also be attractive and stylish in color and cut; it must be sewn elegantly and well. And that is

---

The authors are, respectively, Chief of the Department of Clothing and Knitted-Goods Industry of the USSR Economic Council, a Candidate of Economic Sciences, and a correspondent for Ekonomicheskaia gazeta.

just where the numerous obstacles between production and the consumer make themselves felt.

How can they be avoided? Must the production program of enterprises turning out consumer goods be strictly regulated in all cases as far as quantity and assortment go? Would it not be better to determine it on the basis of direct contracts between industry and trade? This will surely compel the personnel of enterprises and economic councils to make a thorough study of business conditions, of the trend of demand, and to reorganize production effectively.

Under such a system, plants, factories and firms will be working not for the stores, but for the consumer through the stores. And, consequently, many commodities that are available in sufficient quantities can already be excluded from centralized distribution and supplied to the trade network in accordance with orders for them placed by the stores.

Calculations show that the funds spent on improving the quality of goods will be many times less than those now immobilized in old stock that is not in demand.

But how is all this to be done?

It was pointed out at the 5th Session of the USSR Supreme Soviet that in order to have enterprises get a better sense of market conditions and changes in consumer demand, we should engage extensively in the practice of establishing direct ties between enterprises (or associations) and the stores which sell their goods. The time has come to draw up plans for the production of consumer goods on the basis of orders placed by the consumer, with account taken of the establishment of direct ties between the industrial enterprises and trading establishments.

Switching over to such a system will undoubtedly be a step forward, since planning based on orders for goods becomes more specific and is more closely coordinated with the requirements of the population. The economic experiment being carried out at two of the country's clothing firms — the Bol'shevichka in Moscow and the Maiak in Gorky — is in keeping with this task.

We shall discuss this experiment here.

PLANNING AND MATERIAL INCENTIVE

From July 1, 1964, the Bol'shevichka and Maiak associations, as an experiment, were switched over to a new system of planning the production of consumer goods.

These enterprises now draw up their production plans themselves on the basis of orders for goods placed by trading establishments and direct contracts concluded with them.

The heads of the production associations and the directors of
the stores jointly decide on delivery dates and the quantity and
quality of the garments as regards style, model, and color. The
direct contracts also stipulate the prices for the whole range
of goods, as well as the terms of packing, marking, storing, and
transporting the clothing, etc.

On the basis of orders on hand, the firms determine the volume of production and sale of output, the required materials,
and the wage fund.

The firms are also given the right to fix the prices on new
goods and the higher prices for finer finishing. This is how it is
done. The cost of production of the new goods is calculated and
then submitted to the stores for approval. Then an order is
issued within the association for the purpose of approving the
new retail prices and additions to them. Copies of the order are
sent to the stores with which contracts have been concluded.

The goods on which prices and surcharges have been fixed
are given a designation, consisting of a conventional cipher and
an ordinal number. The latter are indicated on the marking
labels.

The production plans that the firms elaborate on the basis of
the orders placed by the stores are submitted to the economic
councils of the economic areas, and also to the Economic Council and the State Planning Committee of the RSFSR. The latter
take them into account in the overall plan of development of the
republic's economy.

Does this undermine the foundations of planning? No, it does
not. In concluding contracts with stores, the enterprises proceed from the need to operate their equipment at full capacity
and ensure high economic indices.

Hence, the experiment that is being carried out not only does
not weaken the planning principle; it strengthens its economic
validity.

The Volume of Goods Sold on Orders and the Percentage of Profitability (Profit) Are the Firm's Report Indices

The experiment also provides for mutual material responsibility on the part of enterprises of the clothing and textile

industries and trading establishments for accurate observance
of the interests of the customer. If a textile mill is just one day
late in delivering fabrics to a clothing factory, it loses 0.3% of
the value of the undersupplied goods. A twenty-day delay will
cost it a forfeit equal to 5% of the value of the undersupplied
goods.

A forfeit of equal amount is paid to the store by the clothing
firms if they fail to meet their delivery dates. Clothing firms
that break contracts by refusing to accept fabrics they have
ordered must pay a forfeit of 5% of the value of the rejected
goods in retail prices.

So it is to the mutual interest of the textile workers, garment
workers, and trade personnel to adhere strictly to the terms
set down in the direct contracts and, hence, to conform to the
interests of the consumer.

The new system of work grants broad rights to firm personnel. They themselves set the norms for stocks of raw materials,
auxiliary materials, and finished goods. If the firms should
need stocks exceeding their own circulating assets, they will be
given credit by the State Bank of the USSR. The main thing is
for the enterprises to base their activities on utilizing raw and
other materials most rationally and on ensuring exact execution
of the orders of the customer.

The wage fund of clothing firms also depends upon them. The
production associations independently determine, on a yearly
and quarterly basis, the volume of production and sale of output, and, hence, the wage fund too. The State Bank pays it out,
of course, at the expense of the funds received by the firms
from the sale of their products. As we see, the economic levers
act in only one direction in the given case too: working at a
loss is excluded!

If a firm fulfills precisely all of the goods delivery commitments to trading organizations it has entered into under direct
contracts, it may transfer 4% of its total profits to the enterprise fund each quarter. The bigger the profit, the larger the
enterprise fund.

And, finally, the firms are given the right to assign funds
from their profits in order to provide incentive for personnel.
Engineers, office workers, technicians, and management personnel are paid only half of the bonus due on the basis of the
results of work for the month under review. They get the other
half when the results for the quarter are totalled. In this way,
financial incentive induces everyone constantly to maintain a
high rhythm of production.

Material incentive is also aimed at achieving goods of high quality. The time-plus-bonus system of remuneration has been introduced for operations that ensure high quality. The rates of a pieceworker are retained here, plus a bonus of up to 40% of the wage rate for excellent work.

Thus, on an experimental basis, a system has been introduced under which a firm has a material interest in producing high quality goods and, on the other hand, answers financially for the results of its activity. In such conditions, firms really work not for the store, but for the consumer through the store.

INTERRELATIONS WITH
STORES AND SUPPLIERS

The Maiak firm's store is situated on Sverdlov Street, a lively thoroughfare in the city of Gorky, and the store of the Bol'she-vichka firm — on Gorky Street in the heart of the capital. They stand out noticeably among the other stores. But that is not what mainly attracts the customers, although, of course, the signifi-cance of the internal and external appearance of a trading es-tablishment cannot be ignored. A big selection of high quality clothing and attention to the customer are above all what have made these stores popular with the public.

The stores linked with the firms reflect the work of these production associations. It is not hard to see here the ad-vantages of direct ties between production and trade, of a firm working for the customer. By way of confirmation, we offer the table on the next page.

The data indicate that at both stores the trend is not only toward an appreciable overfulfillment of plans for the sale of goods, but also toward a drop in leftover stock. As compared with the third quarter of 1963, leftover stock at the Maiak store in the third quarter of 1964 decreased 2.4 times and stock turn-over increased more than 3 times. At the Bol'shevichka store, leftover stock dropped by nearly 20% relative to the norm.

Thus, the goods of the firms do not lie around in the store stockrooms and the funds spent on producing them are not im-mobilized in "non-liquid" goods.

Economists have calculated that if we could hasten the cir-culation of commodities in the country's trading establishments by just one day, the result would be a saving of about 200 mil-lion rubles for the national economy. But the Maiak store, you know, managed to accelerate the circulation of commodities by

Indices	Store of the Bol'shevichka firm*	Store of the Maiak firm	
	3rd quarter of 1964	3rd quarter of 1963	3rd quarter of 1964
Plan (thous. of rubles)	1,380.0	720.0	769.0
Fulfillment (thous. of rubles)	1,437.4	765.5	880.9
Percent of fulfillment	104.2	106.3	114.6
Leftover stock (thous. of rubles)			
a) norm	1,106.0	480.0	367.0
b) actual	886.0	690.0	285.0
Stock turnover (in days)			
a) norm	76.0	61.0	46.0
b) actual	52.0	73.5	23.5

*This store became the firm's store at the beginning of last year.

22.5 days, and the Bol'shevichka — by 24 days. Such are the splendid results turned in by clothing stores operating on the basis of direct contracts with clothing firms.

Store and firm personnel have a common concern, namely, the best possible satisfaction of demand. They make great demands of one another. Firm experts frequently come to the stores and work side by side with the salesclerks. And that is the right thing to do. Not only the salesclerk, but also the supplier has come face to face with the customer, has learned the customer's opinion of his work.

Store personnel, in turn, go to the shops of the production association. There they meet the shop chiefs, foremen, quality inspectors, designers, and stylists. They are very strict in their demands: they want garments without the slightest defect. And such defects still pop up — an uneven stitch or something else of the sort.

One might think that these are trivial details. But no! The customer knows where he is going — to a store selling garments manufactured by well-known production associations. And the firms cannot — they do not have the right to — discredit their trademark because of some "trifle." In this, they act in concert with the store; they have a common interest. And in the final analysis the customer profits.

The clothing manufacturers also established business relations with most of the other stores that concluded direct contracts with the firms. But these relations did not become established of themselves. There were no few misunderstandings in the first months of work. Inveterate habits that had formed in the days when commodities were "distributed" had their effect.

What was the situation like a short while ago? A batch of similar suits or coats would come off the conveyer. They would then be taken to the warehouses of the RSFSR Ministry of Trade's Clothing Administration [Rostorgodezhda], where the personnel far from always received the goods with open arms, for a great many coats of exactly the same style and color were already lying around there for months. But the trade personnel were essentially helpless in the situation. There was a middleman between production and trade — Rostorgodezhda. The stores dealt in whatever it "gave" them; hence the wariness of trade personnel: "suppose we become overstocked?"

Many stores were also very "wary" about placing orders for goods manufactured by the Maiak firm during the second half of last year. An analysis of the orders revealed that, as regards some items, the stores practically ordered one of each. True, in the summary order for all stores, there were from 50 to 600 units of individual models. But after they had been arranged by size, length, type of cloth and color, it was established that it would be very hard practically to fulfill such orders in industrial production — they would require almost individual cutting.

And, unfortunately, the situation as regards orders is the same this year. The Central Department Store, for instance, placed an order for January with the Bol'shevichka firm for 100 model I-183 suits, including 10 of size 48 length 2 and 10 of size 48 length 3. The situation is no better as far as orders for other models are concerned. To fulfill such orders, the firm must have all the types of cloth and all the colors on hand at once, that is, the half-year's stock of them. But the main thing is that consumption of cloth per finished unit increases. Such excessively "small" orders do not promote satisfaction of the public's requirements and do not make it possible to study demand properly.

It cannot be said that all the stores that receive the firms' garments are doing equally good business. Business is much better at the stores which have the best trained staffs of salesclerks, which are attentive to each customer and strive to help him find what he needs, which study demand and advertise best.

Let us take two stores in the city of Dzerzhinsk — Gorprom-

torg's Store No. 10 and the department store — as examples.
The former has many customers; the salesclerks barely manage
to answer their questions and to hand them garments to try on,
and they always strive to select a coat of a style and a color
that will suit the customer. The motto there is: "Not a single
customer must leave without a good purchase." The department
store, on the other hand, has very few customers; the salesclerks are bored and lazily converse with one another. And yet
they wonder why customers do not come to their store.

We also know that at stores which have introduced open display of merchandise, thus facilitating the selection of items,
turnover is greater and there is less stock on hand. But such
conditions, unfortunately, have only been created, in the main,
for the stores of the Bol'shevichka and Maiak firms.

Moreover, conservatism is still a vital force among certain
trade personnel. For instance, in the large Avtozavod District
of Gorky, which has an excellent department store, things have
been so organized that fur coats and cast-iron frying pans are
sold in nearly the same section of the store. The department
store opened up a branch — a store called "Iunost" — a short
distance away, which had been planned as an establishment
specializing in clothes for boys and girls. But even in these
splendid new premises they finally began selling just anything
at all.

Specialized Stores Are Needed to Sell
the Output of the Firms. Only in Such Stores
Is It Possible to Organize the Study of
Consumer Demand and to Satisfy It Properly

Then there will be no misunderstandings in drawing up orders,
the actual requirements of the stores for the firms' goods will
be more correctly taken into account, and the clothing manufacturers will be able to plan production better and to work the
equipment at full capacity.

Under the new conditions of work, there must also be good
relations between the clothing manufacturers and the textile
people. What should they be like?

Textile mills show the clothing manufacturers samples of
new fabrics seven months before the beginning of the year. Then
direct contracts are concluded which specify the designation,
types, sorts, designs, colors, prices, quarterly consignments,
and the total order for the year.

The Clothing Firms Themselves
Pick Their Suppliers of Fabrics

The clothing firms can order fabrics from any textile mill in the country.

Provision is also made for the possibility of changing orders for fabrics, and also, incidentally, for the garments manufactured by the clothing firms. In such cases the client must inform the supplier no later than two months before the beginning of the quarter.

But much still has to be done to improve relations between the clothing manufacturers and the textile firms.

Forty-nine textile mills supply the Maiak firm with cloth; 32 of them supply woolen fabrics. The principal suppliers are the Thaelmann Textile Combine in Leningrad, the Krasnodarsk Worsted Cloth Combine, the Osvobozhdennyi Trud Factory in Moscow, the First of December Factory in Estonia, the Krasnye Tekstil'shchiki Cloth Factory in Sumy, the Lenin, Sverdlov and Razin factories in Ul'ianovsk, the Odessa Cloth Factory, and the Krasnyi Oktiabr' Factory in Penza, among others.

The Bol'shevichka firm has 39 suppliers. Of these, 13 provide it with woolen cloth, 14 — with auxiliary materials, and 12 — with accessories. Numbered among the suppliers are the Kuntsevo Factory and the Peter Alekseev, Minsk, and Briansk combines.

The clothing manufacturers selected the fabrics for work under the new system at the Inter-Republic Fair in Moscow from samples supplied by the textile mills themselves, with the color for each item of cloth indicated. It should be noted that they first selected the fabrics and then took the orders for the models of clothing from the stores, rather than doing it the other way around. But the clothing manufacturers did not have the time to make lengthier preparations for working under the new system, and they were faced with the fact of either ordering the fabrics or remaining without them.

However, it proved far harder to get the new materials than to look them over at the fair. The firms were to have concluded direct contracts with the enterprises before July 1, 1964, for the delivery of fabrics, accessories, and fur collars in accordance with the orders placed by stores for the second half of 1964. The USSR Economic Council gave all enterprises instructions to this effect. But some of them long and persistently avoided concluding contracts (with the Gorky firm, for instance).

Strange as it may seem, the Osvobozhdennyi Trud Factory in
Moscow, the Iskozh Combine in Kalinin, and the Trud Fur Factory Association in Moscow were among the "refusers."

But that was not the end of the trouble. Some factories sent
the firm fabric colors that it had not ordered, attributing this
to a lack of dyes. An Ul'ianovsk factory, for instance, instead
of delivering bright, rich fabrics designed for children's coats
and specified in the direct contract, sent the Gorky clothing
manufacturers fabrics of dirty blue and dark green that were
absolutely unsuitable for children's clothing. Naturally, all of
them, and there proved to be no more nor less than 1,960
meters of cloth, were returned to the unscrupulous supplier.
As a result, the firm failed to supply the trading establishments
with many children's coats. The clothing manufacturers had to
spend time replacing the missing fabrics so as to keep the conveyer going. And this, in turn, led to the substitution of fabrics
and styles and to the violation of the terms of the contracts with
the stores, for which the firm had to pay the stores a fine of 5%
of the total value of the goods delivered.

The following enterprises, among others, are noted for their
failure to fulfill commitments in a precise manner: the Krasnyi
Oktiabr' Factory in Penza, the Odessa Cloth Factory, the Thaelmann Combine in Leningrad, the Parizhskaia Kommuna Factory
in Riga, the Minsk and the Briansk combines, and the Peter
Alekseev Factory. They are late in making deliveries and send
low-grade fabrics and colors that were not specified in the
contracts.

Poorly finished and even clearly defective fabrics cause the
clothing manufacturers much grief. For instance, the Maiak
firm returned to the Ul'ianovsk Factory cloth which was subject
to excessive shrinkage (up to 7%). Variously shaded and low-
quality fabrics were returned to the Thaelmann Combine in
Leningrad and the Alma Ata Cloth Factory. It is clear that the
textile manufacturers are continuing to work in the old way, that
they are sending off to the clothing manufacturers everything
they have in their warehouses. At the fair in Moscow they ex-
hibited fabrics with rich, bright colors, but they are supplying
the clothing enterprises with colors that have nothing in common
with the samples, although they mark them as the same colors.
It is true that the firms, taking advantage of the economic sanctions provided in the direct contracts with the suppliers, regularly receive a forfeit from them, but that is no help to the customer. And under the new system, the relations between the

textile and clothing manufacturers must be built on an entirely different foundation!

To dress our women beautifully, we must also reject monotony in trimmings. Take fur collars, for instance. The Kazan Fur Combine and the Belka Fur Factory of the Volga-Viatsk Economic Council supply them to the firm in practically one color only — black — and the collars are of poor quality. The Trud Fur Factory Association in Moscow refuses to supply collars of grey karakul and kolinsky. The autumn hunting season is long over, and yet the Belka Factory has not sent to Gorky a single red fox; coats manufactured by the firm have been lying around in the warehouse for a long time without collars. The Briansk Combine was to have supplied the Bol'shevichka firm with 3,000 meters of fabric in November of last year, but it has not sent a single meter as yet.

Many misunderstandings have arisen between the clothing manufacturers and the suppliers of auxiliary materials. The materials sent are often of poor quality. Recently, for example, the Alatyr Factory in the Chuvash ASSR sent off 5,500 meters of damaged sheet wadding to Gorky, and the Kozhzamenitel' Factory in Bogorodsk did this firm "a good turn" by sending it 4,500 meters of defective pavinol [?].

The enterprises we have named are situated in different parts of the country. Obviously, the economic councils of the economic areas and also the economic councils of the RSFSR and the USSR will have to take measures to eliminate the above-mentioned shortcomings in supplying the clothing firms with fabrics and accessories.

ORGANIZATION OF PRODUCTION AND WORK

The clothing manufacturers realize how much they must still do in order to have the Soviet people fully satisfied with their products. But how are the technological processes that have taken shape over the years to be adapted to the new system of work? How should production and work be organized so that the enterprise can switch over at short notice to producing the assortment of goods that the stores need, while ensuring top quality at the same time?

Before the transition to the new system, large-scale technological lines (processes) predominated in the main at the garment factories of the firms. It is not easy, of course, to control such processes, especially in sewing suits and women's clothes, since their styles and colors change too frequently. A foreman often does not have enough time to check on the fulfillment of each operation, to tell a worker how to correct one or another defect. And this leads to shortcomings in the quality of the output.

It stands to reason that under the new conditions of work, the production of big batches of goods, running into thousands of units of one type, is out of the question, inasmuch as the stores cannot and will not place orders with the firms for as large a number of models of suits and coats as they used to. Under the old system of organization of the processes, where the conveyers turned out 120 to 150 units per shift, production would have to be reorganized nearly every day to execute such orders! Furthermore, it was found that requests for top-quality goods predominated in the orders placed by the stores.

After studying the experience of the best clothing factories in our country and the fraternal socialist countries, the clothing manufacturers came to the conclusion that the quality of output greatly depends upon the length of the conveyer. And that is quite understandable, for small production lines are easier to reorganize, to adapt to the demands of trade.

Here is what was done at the Maiak firm. The long conveyers were broken down into separate technological sections, the number of workers on each depending upon the initial production areas. Thus, two processes were singled out from the former multi-style sectional Conveyer No. 1. Production of women's top-quality coats was organized at the preparatory section of the conveyer, and the process of sewing women's coats of better quality — at the assembly and finishing section.

Former Conveyer No. 2 was also divided into two sections. A small production line for sewing top-quality clothing was set up at its preparatory section, which, in addition, was physically detached from the conveyer.

At the other sectional conveyer, which now was also divided into two processes, the garment people organized the production of small batches of goods out of synthetic and chemical materials. At the assembly and finishing sections of this conveyer, the Gorky garment workers are now sewing women's coats that are in great demand.

Much has been done to improve the technological process at the Bol'shevichka firm as well.

Reorganization of Production Has Affected
All the Technological Lines and Processes
in the Shops of the Associations

The very nature of the technology and organization of labor has changed. The enterprises switched over from mass-scale flow production to small-batch production in a very limited period of time.

Naturally, all this could not have been done without material outlays. To switch over from big conveyers, where each worker specialized in one particular operation, to small production lines, where each worker performs several operations, it was necessary not only to retrain people, but also to increase the time norms for working the articles.

The garment workers had been "stuck" for years on one and the same operations. The large conveyer had limited their skill. Now horizons are expanding and they must know more. The firms organized mass training of personnel.

In the course of the reorganization of production, much was done to study all the operations of the old and newly organized processes. The task was to reveal the slightest possibility of improving the quality of each operation, to raise the skill of the workers to the maximum. From the very first days of operation of the new system, the garment workers had their attention focussed on a single requirement: high quality.

As a result, firm specialists, production innovators, and rationalizers submitted many valuable suggestions aimed at perfecting methods of processing articles, raising the quality of goods, and improving the organization of production. And it turned out that the previously accepted technology for manufacturing output of better quality does not at all meet the rising demands of the consumer. Therefore, the sewing of suits and women's garments on the basis of this technology was in many respects brought closer to the technology for manufacturing top-quality goods. This was done by introducing additional operations — additional finishing of certain parts in the sewing shops and additional pressing — without which it was hard to count on producing first-class goods. It is interesting to note that the ironing of such pieces as the fronts and coat-breast interfacings had long been done on presses with protuberant forms. And now it turned out that this method by no means ensures high-quality fulfillment of the operation. Irons had to be used.

Nor could certain other costs be avoided. These included, first and foremost, some additional hand work (bar and other tacks, tacking of facings and the upper collar, and working expensive collars) and additional time for final pressing and finishing of the goods. Sewing suits and coats is delicate work. Sometimes only individual skill, raised to the level of art, can ensure excellent execution here. And the managers and specialists of the country's clothing enterprises should not be disturbed about the production associations' experiment with more hand operations. This experiment was economically justified. The excellent merchandise has suited the customers' taste.

Still another important innovation was introduced at the firms: interoperational inspection, which permits assessment of the correctness of execution of one or another operation.

It must be confessed that the personnel of the technical control divisions have shut their eyes in the recent past to the output of low-quality goods. Such an approach to the matter was made legitimate by the very system of material incentive for the control personnel, who could receive a bonus only if the entire section fulfilled the production plan. And the foreman frequently utilized this to "bring pressure to bear" on the control personnel, saying: "Why have you stopped accepting output? We are fulfilling the plan!"

Now the clothing manufacturers have revised the regulations governing material incentives for technical control personnel, who today immediately stop accepting goods if they discover a defect.

But making excellent garments is still not the whole of it. They must be kept in marketable shape. This is done by sending the garments down to the stockrooms for finished products on hangers, where they are suspended on racks, and also by delivering them to the stores on hangers in special vans. The firms plan to mechanize the loading and unloading of finished goods so that they will go directly from the shop to the customer without once having to be bundled onto the shoulders of a loader.

ECONOMIC INDICATORS

The Bol'shevichka Clothing Association is a large specialized enterprise. The association's annual output is worth more than 50 million rubles. The other firm, the Maiak, is less specialized than the Moscow firm and produces many types of garments. Its annual output is worth about 40 million rubles.

It was not by chance that these two associations were selected. The Bol'shevichka makes men's suits and therefore requires less diversified fabrics. The Maiak, on the other hand, works with a wide assortment of women's garments and fabrics. It must deal with a very large number of suppliers, which, naturally, complicates execution of the orders placed by trading establishments.

But the associations were switched over to the new system of planning and selling output without giving them sufficient time to make preparations and conclude contracts with stores and suppliers. Hence, they found themselves hard pressed at first. The stores demanded the shipment of goods in the assortments they had ordered, but this could not be done because the textile mills were reorganizing production. Hence, the rhythm of work of the firms was upset.

All this, naturally, could not but tell on the economic indicators of the work of the associations in the new conditions. Average daily output in retail prices amounted to the following (see first table below, in thousands of rubles).

It is evident from the table that the Bol'shevichka firm had not quite reached the average daily output achieved before the

Firms	Before the switch to direct contacts (1st half of 1964)	After the switch to direct contacts				
		July	August	September	October	November
Bol'shevichka	182.2	130.1	130.3	173.4	177.8	179.0
Maiak	106.2	74.4	83.8	107.1	112.0	108.5

Firm	Average per month in the first half of 1964	July	August	September	October	November
Bol'shevichka	276.8	209.6	231.3	266.5	307.3	278.8
Maiak	525.2	381.2	462.8	494.0	485.0	481.0

switch to the new system. The value of the commodity output also dropped somewhat because the average price of a suit in the first half of the year was 87 rubles, while it was only 81 rubles in the second half.

The Maiak Association had already topped the average daily commodity output attained in the first half of the year by more than 2%.

The main result of the second half-year of work was that all the firms' output was fully sold and in greater demand. The following data testify to this. Whereas the Bol'shevichka's plan for the sale of finished goods was overfulfilled by 0.9% in the first half of the year, it was topped by 14.5% in November. The corresponding figures for the Maiak firm are 0.7 and 11.1%.

It should be noted that in the first half of the year the output of the Bol'shevichka firm was sold in 230 stores, and that of the Maiak firm in 228, while in the second half each firm sold its goods in 22 stores.

And what the second indicator of the work of these enterprises — the profit (in thousands of rubles) — looks like is shown in the second table on page 245.

Notwithstanding the substantial reorganization of production and considerable outlays in these firms, there was a gradual growth of profit as compared with July. But the absolute amount of profit for the considered months was somewhat less than in the first half of the year. However, this decrease is not connected with the work of these associations.

The point is that a different percentage of profitability has been fixed for almost identical items in the price lists. For instance, the retail price for a man's suit in Price List No. 15 provides for a profitability of 7.5%, and for a suit of the same cloth in Price List No. 10 — a loss of 2.4%. And if the store wants an assortment of goods with a smaller percentage of profitability in the retail prices, the size of the profit will naturally decrease regardless of what the producer does. The Price Bureau of the USSR State Planning Committee has not yet eliminated this serious shortcoming.

Nor does the loss of profit from a cut in the average price on an ordered suit depend upon the enterprises. We think that there is absolutely no justification for the tendency of trading establishments to go after cheap goods only. Even now the stores are not always satisfying the demand for goods of medium and higher price, that is, goods of higher quality, made of better fabrics.

Labor productivity at the firms is calculated on the basis of the normative cost of processing the goods. An analysis of data on average output shows that its rates of growth are not high at either the Bol'shevichka or the Maiak. Introduction of all-round mechanization should help the firms raise labor productivity.

An important condition for development of the economy of the enterprise is a growth in labor productivity that outstrips the growth of wages. Only in such a case can the enterprise achieve a systematic drop in its production costs and an increase in accumulations. Let us see how this condition is observed at the firms under consideration.

Period	Coefficient of comparison of labor productivity with wages	
	Bol'shevichka	Maiak
First half of 1964	1.046	1.498
July	0.903	0.916
August	0.964	0.888
September	1.193	1.050
October	1.090	0.975
November	1.086	0.921

In the Bol'shevichka Association, output per worker, if the vacation months (July and August) are not taken into account, outstrips the growth of wages. The comparison coefficient for this enterprise is greater than unity.

The situation is not too good in the Maiak Association. Wages outstrip the growth of labor productivity for nearly all the months it was working under the new system. The firms' economists will have to analyze the causes of this phenomenon and map out ways of establishing a correct relationship between the growth of labor productivity and wages.

THE MAIN CONCLUSION

The Economic Experiment of These Associations Opens Up Extensive Possibilities for Improving the Work of All Our Industries Producing Consumer Goods

Direct ties between enterprises and stores make it possible to draw up plans for the production of consumer goods on the basis of orders placed by the consumers. Hence, planning becomes more specific and is more closely coordinated with the requirements of the economy and the population.

Planning production on the basis of orders placed by the consumers promotes a sharp rise in the quality of output. The garments manufactured by the Maiak and Bol'shevichka firms no longer lie around in the stock rooms and warehouses as dead capital, and the national economy gains directly from this, for the country loses millions of rubles each year from price reductions on "unmarketable" commodities. In addition, there is a sharp curtailment in the periods of commodity turn-over, which also results in a real saving for the national economy.

The new system is also advantageous to the personnel of the enterprises. Planning on the basis of orders placed by the stores extends the economic independence of the enterprises and increases their responsibility for selecting the most economical ways of fulfilling assignments. Under these conditions, the size of the enterprise fund and financial rewards for personnel depend to a greater extent upon the personnel themselves.

The following conclusions also suggest themselves.

First, in the future it would be advisable to give an enterprise with a complex and wide assortment of goods a longer period in which to prepare for the switch to the new system of work. In the case of the Maiak and Bol'shevichka firms, the period was essentially confined to two months. The most propitious period would be six months.

Second, the USSR Economic Council did not provide time for the accumulation of finished goods in the firms' stock rooms against orders placed by stores. And so it turned out that on July 1 the stores were already demanding garments from the firms, whereas the latter had barely put them into production, had not accumulated the necessary quantities, and, hence, could not make up consignments as ordered and deliver the goods to the stores. This means that enterprises following the lead of the Bol'shevichka and Maiak firms will have to assign a minimum of one month especially for the purpose of building up the stock of finished goods.

Third, it has become obvious that a big clothing association cannot work with small stores. It is expedient to have no more than five or six big stores selling the output of a local clothing association on the basis of direct contracts. Many of the difficulties encountered by the firms during the first months were due precisely to the fact that they had to deal for the most part with small stores.

The operation of firms on the basis of orders placed by the consumers urgently raises the question of extending the economic independence of the enterprise and the rights of its director. Associations should be given the right to fill orders placed by consumers in other towns, to send them the goods they request on an individual basis. This requires a special mail-order department.

To fill the individual orders of buyers for so-called non-standard garments regularly and in good time, the firms must have special salons that are supplied with the necessary quantity of semimanufactured goods.

The schedule for the execution of contracts with stores has the force of law. Local organizations, guided by "regional" needs and considerations, must not be permitted to reshuffle the schedule at their discretion, thus putting the firms in a difficult position with respect to their customers.

The USSR Council of Ministers approved the proposals to expand the experiment of the Bol'shevichka and Maiak associations and charged the USSR Economic Council with expanding the range of enterprises that have been switched over to the new system of planning and selling output. It is a matter of honor for the enterprises and organizations participating in this major state undertaking to reorganize their work so that all their output is of high quality and fully satisfies the demands of the population.

Pravda, January 17, 1965

An Important Economic Problem

N. Fedorenko

In recent years the Soviet Union has scored major successes in developing the most modern branches of industry, such as chemicals, electronics, and machine building. The 1964 plan has been successfully fulfilled. This year the national income, utilized for consumption and accumulation, will rise by 8%, and the gross output of industry — by 8.1%.

All this demonstrates again and again the advantages of the planned socialist system, which assures continuous, dependable, and stable development of the economy. However, while emphasizing the advantages of our system, we clearly understand that it contains vast unutilized possibilities, the rational employment of which would give it new strength with which to improve the rates of growth of production and the level of material well-being of the working people. This requires, above all, improvement in the quality of planning and management of the economy. The economic difficulties that have manifested themselves in a number of instances — inadequate growth rates in particular branches, temporary disproportions in their development, and difficulties in supply and marketing — have one common cause: the fact that the management of the economy does not correspond to the level of development of the productive forces.

New economic conditions have now come into being which demand new solutions in the area of economic administration. Acute shortages in production are disappearing. Renewal of the output assortment and improvement of its quality are increasingly real developments. The "theory" that production must proceed regardless of cost is passing into history. Stable direct connections between consumers and producers are being

---

The author is a Member of the USSR Academy of Sciences.

established on the basis of a vast growth in the number of enterprises and changes in their geographic location. Our enterprises have become stronger in the financial and economic sense, and do not need excessive regulation and tutelage.

Under these conditions the old forms of management and planning must be replaced by new ones. This is the objective need of the day. The more rapidly it is met, the faster we will move ahead.

Rational employment of all resources — labor, natural, and technical — is a guarantee of the successful fulfillment of the great tasks which the party has placed before the economy.

The direction to be followed in this effort is further development of the initiative of the masses of the people through skillful employment of the economic levers of management in the hands of the socialist state. Articles recently published in Pravda and other national newspapers and journals contain a number of valuable proposals with respect to the further improvement of such economic levers as material incentives for production personnel, the strengthening of cost accounting at enterprises and production combinations and the expansion of their rights, planned price formation, financing and extension of credits, the development of direct connections among enterprises, the strengthening of ties between industry and its customers, etc.

Individually, none of these levers is capable of resolving the problem of mobilization of all resources. But together, in interaction and as a totality, they can be coordinated so as to provide the best possible guidance to the initiative of the masses toward fulfillment of the assignments of the state plan.

A system of optimum prices aimed at providing planned incentive will have no serious significance, for example, if the plans of the enterprises are "handed down" on the basis of previous indices. The economic optimization of truck transport shipments will not achieve its objectives if the earnings of truckdrivers and engineering and technical personnel of trucking enterprises continue to be calculated in the old manner, by ton-kilometers.

Thus, proposals to improve and make wider use of various economic levers of management must be considered in toto; there is need for detailed study not only of the operation of each of them individually, but of their interaction.

Skillful employment of the entire system of economic levers will doubtless yield major results for the economy.

However, this is not enough. This source for improvement of the effectiveness of social production must be supplemented by another, which is also highly important. We speak of improving the quality of economic planning and centralized management of the process of socialist reproduction.

While making broad and skillful use of economic levers and market mechanisms, we must never forget that centralized, unified economic planning is one of the greatest accomplishments of the socialist system.

There is no disputing the fact that excessive detail in plan assignments, the regulation of everything, is an evil that leads to irrational expenditures and interferes with flexible, efficient adoption of new and progressive solutions. It came into being against a background of impaired material incentives and is inevitably retreating into the past as these incentives increase. But the vast, highly specialized modern enterprises working for the entire national market, and even the world market, require centralized accounting of their needs and allocation of orders. Without this it is impossible to organize rational direct connections, impossible to assure priority satisfaction of the needs of the most important branches.

It is necessary not to weaken but to improve centralized planning. Only in this way is it possible to assure optimal rates of development of the leading branches, to balance proportions, to redistribute capital investments, to establish proper relationships between accumulation and consumption, between consumption paid for out of personal incomes and that paid for out of social consumption funds, and to determine price and wage policy.

The economic levers under discussion serve to guide working people toward fulfilling the tasks of the economic plan, and whether their labor will be fully utilized to bring about a further upsurge in the economy and the people's well-being or will in some part go for naught is entirely determined by whether the economic plan is optimal for the given level of development of the productive forces.

These two problems — utilization of economic levers and incentive methods, on the one hand, and organization of optimal planning, on the other — are intimately and indissolubly intertwined. They cannot be torn apart or counterposed to each other. They are two inseparable aspects of the process of further development and reinforcement of democratic centralism in the administration of our economy.

There are urgent measures that must be taken right now to eliminate shortcomings in the organization of management of the economy. But it is very important not to repeat the past and, in the future, not to confine ourselves to halfway measures.

For the purpose of consistent improvement of the system of planning and management of the economy of the USSR, we need a clear and precise long-range outlook, exactly as in the field of economic development. And we are already in a position to formulate what this is: the establishment of a single system of optimal planning and management of the economy on the basis of extensive utilization of the techniques of mathematical economics and computers. This goal is entirely attainable under our conditions, in a country with a planned economy, highly developed productive forces, and advanced science and technology.

We note at least three most important components of a unified, automated system of optimal management and planning: an interrelated complex of models of economic processes and phenomena, a single system of economic information, and a unified network of state computer centers.

Karl Marx wrote that a science becomes an exact science in the full sense of the term only when it possesses its own mathematical apparatus. A synthesis of the achievements of economics, mathematics, cybernetics, and computer technology has produced the most advanced trend in economics — mathematical-economic methods. Profound study of objective economic laws makes possible the creation of mathematical models, in which technological and economic connections within the economy are described with precision and brevity. With the assistance of these models, constituting in essence no more than exact descriptions of carefully studied phenomena, the personnel engaged in economic planning can adopt well-founded decisions on the basis of calculations of numerous variants, consider complex economic interrelations, and examine the remote consequences of each economic measure or scientific-technological discovery.

It is also extremely important that the creation of mathematical models is the chief prerequisite for the introduction of computers into all spheres of management of the economy. Computers cannot function on the basis of indefinite verbal descriptions; they require the exact mathematical formulation of every problem. Only when this is done is it possible to perform with their aid, in a matter of minutes, calculations that would require many years if ordinary methods are used.

In order to achieve scientifically valid planning and management, we must have mathematical models of the economic life of enterprises, of combinations of them, of branches, and of the economies of districts, republics, and the country as a whole. These and other models must be rigorously dovetailed.

Such already well-developed mathematical models as the interbranch balance make it possible to construct fully balanced plans for the development of the economy, with allowance for the most complicated ties among branches. This is already finding effective application in the practical work of the planning agencies. However, we cannot be satisfied now merely with a balanced plan. In each concrete case we are obligated to seek and find an optimal solution, i.e., the one best solution for the given conditions: maximal employment of productive capacities and other resources, the best interbranch proportions and variants of geographic distribution of productive capacities or freight traffic, and the optimal trend in capital investments that secures the highest possible return.

The methods of mathematical programming, and various other mathematical methods, assure us of rapid production of optimal solutions for each specific planning problem encountered in the management of the economy. Techniques of optimal management and planning must be applied both in the country's economy as a whole and in each of its components: enterprises, branches, republics. The establishment of an interrelated complex of mathematical-economic models and mathematical means of processing them, and of optimal solution of problems of economic planning, constitutes at the present stage <u>one of the most important tasks of the Soviet science of economics.</u>

However, a mathematical-economic model is merely the framework for the solution of a problem. Its real content is determined by <u>economic information,</u> i.e., by concrete real data on capacities and resources, on the shipment of output, its distribution and consumption. According to modern scientific concepts, any process of management is in some sense a process of treating information. The quality of a plan and the correctness and effectiveness of orders are wholly dependent upon the accuracy, reality and timeliness of the information at our disposal in drafting plans and making decisions.

The management of our economy requires the processing of gigantic quantities of economic information. These flows of information will no longer fit into the outdated forms that have come into being historically on the basis of manual methods of management. Of the broad complex of problems involved in further improvement of the system of planning and managing the economy, the task of providing information for economic planning is one of the most immediate and important. The contradiction between the needs of exact and flexible planning and management, on the one hand, and old, backward forms of document circulation and communication, on the other, is manifested with particular acuteness in this field. It is perfectly obvious that the data contained in the current system of statistical reporting are unsuited to these new objectives.

With the present organization of documentation and document circulation, computers can yield no benefits because the mere preparation of data for computer input now takes considerably more time and means than planning calculations that are performed manually. Thus, a fundamental rationalization of the system of economic information is an essential prerequisite for efficient introduction of computers.

Improvement of the information system has to begin directly at the point of production, in the shops and enterprises, and must embrace all levels of planning and management. The information system must be comprehensive, uniform, and integrated. Integration means that all forms of information — technological, economic, accounting, financing, supply — must be in compatible forms. The program for accounting is reduced several-fold in a uniform, automated information system. At the same time, the coefficient of effective utilization of the information increases many-fold. Millions of people will be freed from the unproductive and exhausting work of endlessly copying documents, and their time will be made available for creative labor.

A unified state network of computer centers should be the technical base for the unified, automated system of optimal management and economic planning for the USSR. Under the conditions of the highly developed and complex, rapidly advancing economy of the Soviet Union, the introduction of computer equipment into planning and management is essential to assure the most efficient development of the economy. The amount of economic information requiring transmission and processing and the complexity of multiple-variant plan calculations are already so great that even the large number of personnel in the field of economic management cannot effectively perform their duties in the absence of modern computers.

Utilization of the modern scientific mathematical-economic methods and computers is an important aspect of the economic competition between socialism and capitalism. It must not be forgotten that, in the United States alone, there are now some 18,000 computers and computer systems, and that about 80% of their operational time is spent on economic calculations and the processing of economic data. If we consider that the volume of economic work increases approximately as the square of the growth of the number of enterprises and the number of different items produced in the economy, it is clear that by 1980 even the employment of 100,000,000 persons in the machinery of economic administration could not cope with the processing of economic information by manual means. It is only on the basis of universal introduction of computers and systems performing hundreds of thousands or millions of operations per second that it is possible to bring about optimal management of the economy of the USSR.

Under the conditions of a planned socialist economy, it appears possible, if efforts are concentrated on this, to overtake and surpass the capitalist countries in a short period in yet another field — that of effective utilization of computers in the economy. In the capitalist countries, the introduction of computers is taking place within the framework of competition among the monopolies, and this powerful equipment is not effectively employed when considered in terms of countries as a whole. We are capable of bringing about comprehensive automation of planning calculations on the scale of the economy as a whole with a considerably smaller number of computers but a considerably higher coefficient of effective utilization of them. Under the conditions of a planned economy, the possibility exists of planning and establishing a unified state network of computer centers, functioning in accordance with a single system and as one gigantic mechanism.

This network would constitute an aggregate of centers of various sizes, equipped with up-to-date computers; they would be distributed throughout the country and interact through a single automated communications system. The network of computer centers would perform the functions of collecting, transmitting, and processing all economic planning and accounting information, and of calculating optimal economic plans for individual branches, republics, production combinations, transportation, and financial agencies. The unified system of computer centers would include several levels, each of which would serve the economic planning agencies of the given level: the country as a whole, republics and branches, geographic and production complexes. It would rest upon automated and mechanized systems of information and management of enterprises and combinations of them; these systems are already being actively developed in various regions and districts.

The establishment of a system of economic management that rationally combines centralized planning and administration with broad employment of economic levers and functions on a new and modern technological foundation constitutes an unusual task, both in its significance and complexity, even for the Soviet scale of doing things. It includes the development of a scientifically substantiated system of economic measures and the establishment of a new branch of industry — a cybernetics industry — plus a sharp leap in the development of a number of branches of science, and the creation of completely new scientific disciplines.

The objective posed is exceedingly complex but entirely attainable. The Soviet people, led by the Communist Party and employing the vast advantages of a planned socialist economy, will certainly be able to realize this objective.

Much has already been done and is being done to introduce scientific methods and computers into management and planning practice. Work is being conducted at dozens of enterprises to employ computers for calculations in economic planning. A number of economic councils have designed and are establishing automated systems for processing economic data. Definite advances have been made in developing interbranch balances and solving various optimal tasks in planning and managing particular components of the economy, the geographic allocation of production, and the organization and management of construction. In a number of research institutes, such as the Central Institute for Mathematical Economics of the USSR Academy of Sciences, the Central Institute for Business Equipment, the Cybernetics Institute of the Ukrainian Academy of Sciences and others, certain types of mathematical-economic models have been developed, as well as methods of solving problems in production planning, the foundation of a theory of economic information, and methods of designing automated systems of management. The universal introduction of the mathematical methods and models already developed is equivalent for the economy to finding and investing many more millions of rubles.

Our research in the field of economic cybernetics and mathematical-economic methods is progressing successfully. However, the level at which it is conducted is still inadequate to the tasks we face.

The development of a plan for a unified and automated system of planning and management based upon a single network of computers is an extremely complex and responsible undertaking. Major investments of capital for scientific purposes are required. It is necessary to train personnel on a mass scale, to expand the network of scientific and engineering firms and designing institutes by all possible means. All these outlays will pay for themselves rapidly and will yield vast economic dividends.

It would be desirable to proceed without delay to the creation of a single automated system of optimal planning and management of the economy, but to carry out this most important measure by stages. As a result, economic yields greatly exceeding the funds expended upon all the work involved would be produced even as the system develops. The advantage of a planned economy lies in the fact that we are in a position to develop and realize this system on a single, centralized basis. This means that the government agency heading and effectuating the entire complex of work leading to the achievement of this major goal must possess the requisite authority.

Ekonomicheskaia gazeta, January 20, 1965

Direct Contacts Are Expanding

Our readers have asked in their letters how the enterprises of light industry will shift to direct contacts with trade establishments. Here is our answer to this question.

***

It was noted at the 5th Session of the USSR Supreme Soviet that in order for enterprises to get a better sense of market conditions and changes in consumer demand, direct ties should be established on an extensive basis between enterprises (or associations) and the shops that sell their goods to the population.

The Economic Council of the USSR recently adopted a decision under which a substantial number of enterprises of light industry will shift, in 1965, to direct contacts with the stores.

It is contemplated that 128 enterprises will make the shift in the second quarter of this year, including 19 cotton textile mills, 49 woolen textile mills, 8 silk mills, 7 linen mills, 25 tanneries, 6 synthetic leather factories, 10 factories producing accessories, and 4 fur factories.

In the third quarter, all the clothing and footwear enterprises of the Moscow City Economic Council will convert to the production of finished goods in accordance with orders placed by shops and wholesale and retail trade organizations. The new system will also be adopted by the garment and footwear factories of Leningrad, Kiev, Odessa, Kharkov, L'vov, Minsk, Vilnius, Riga and Tallinn, as well as by some of the enterprises of the Moscow (regional), Belorussian, Alma-Ata, Georgian, Moldavian, Armenian, Uzbek, Tajik, Kirghiz, and Turkmenian economic councils.

Thus, in the current year 25% of the clothing factories, 28% of the footwear factories, 18% of the textile mills, and 30% of the tanneries will begin to work in accordance with orders received directly from the shops and wholesale and retail trade organizations. Under these circumstances, special attention must be paid to raising the economic efficiency of the trade bodies and shops, as well as to improving the study of consumer demand.

The enterprises that will enter into direct contracts with the trade bodies are being granted extensive powers. The directors of clothing and footwear factories will establish independently their annual, quarterly and monthly plans with respect to volume of output, labor, costs of production, and other economic indices, in accordance with the orders of the wholesale and retail trade bodies and shops.

The directors of the textile mills, tanneries, and other associated enterprises will determine these indices on the basis of orders received from the clothing and footwear enterprises, and trade enterprises and organizations. In drawing up their production plans, both the first and the second groups of enterprises will base their estimates on the maximum utilization of their output capacity, available raw and other materials, and specialization.

The production and managerial efficiency of the enterprises will be evaluated on the basis of the following system of indices: a) for clothing and footwear factories — fulfillment of the plans for marketing the goods put out in compliance with orders received from trade bodies, profit; b) for textile mills, tanneries and synthetic leather factories, and other associated industries — fulfillment of plans for the delivery of goods in compliance with orders received from clothing and footwear factories and wholesale bodies, profit.

The output plans will be submitted to the respective economic councils of the economic districts.

To meet consumer demand, the shops and wholesale and retail trade bodies will be able to change their orders two months before the beginning of the quarter with respect to the type, model and color of items, and six months before the quarter begins with respect to the texture of fabrics and shoe fashions. The clothing and footwear factories will amend the orders they have placed with the supplier-enterprises 45 days in advance of the quarter.

The managements of the enterprises will amend their plans accordingly and notify their respective economic councils of the changes. The latter will report on the changes in the plan indices to the State Planning Committee of the union republic, and in the case of the Russian Federation, the Ukraine and Kazakhstan — also to the economic council of the republic.

And what is to be done about the reductions in profits stemming from the changes in orders? These losses will be covered by the reserve funds of the economic council of the republic or economic area concerned. (The reserve fund will be built up from the allocation of 5% of the profits derived by the enterprises.)

The light industry enterprises that shift to the new operating procedure will have the right to establish the norms for inventories of raw and auxiliary materials, unfinished output, and finished goods. In doing so, their point of departure will be the rational utilization of the materials and goods and the fulfillment of orders placed by the clothing and footwear factories, shops and trade bodies. If these enterprises build up above-norm stocks of raw and other materials, they will receive credit from the State Bank of the USSR.

The size of the wage fund will also be determined by the enterprise. The State Bank will issue the sums needed to cover the wages in accordance with the wage fund established by the enterprise and the fulfillment of the volume of output expressed in the normative cost of processing. If the plan has been overfulfilled with respect to volume of output, the enterprise will receive additional sums to be paid out as wages over and above the established wage fund: 0.9% of the fund for each 1.0% of plan overfulfillment.

If the marketing and profits plans are fulfilled, the enterprise will be permitted to assign 4% of its actual profit to the enterprise fund every quarter. The director has also been authorized to allocate, for the granting of bonuses to engineers, technicians, office employees and management personnel, a sum from profits that does not exceed 50% of the wages fixed for this personnel under the wage scales.

Workers performing operations that ensure high quality of produce may be transferred from piecework to time-and-bonus remuneration. While on these operations, they stay at the wage rate established for pieceworkers and receive a bonus of up to 40% of the basic rate for high-quality output.

To secure higher quality of produce and better organization of production, the factories have been given the right to create the jobs of chief designer-deputy, chief engineer and sales manager-deputy director and to establish the required staffs for marketing and commercial work. The factories, with the consent of the ministries of trade of the union republics, will open firm shops.

Some trade personnel have been of the opinion recently that it is necessary to order cheap goods at the factories. As a result, the shops are not always able to meet the demand for medium - and high-priced goods, i.e., for high-quality clothes made of finer materials.

In 1965, consequently, it is intended to organize special teams at the Thaelmann Clothing Factory and Clothing Factory No. 9 for filling orders of trade bodies for the latest models of ladies' and men's overcoats in expensive cloths. The quality of these coats will not be inferior to any of the best domestic or foreign models. The directors of these factories have been given the right to transfer the workers of these teams to time work and to grant them bonuses of up to 40% of their wages. If necessary, the directors can raise the workers' basic rates by 15% and the foremen's wages by 20%.

Material incentive is accompanied by material responsibility on the part of the enterprise for the fulfillment of orders. For each day's delay in the delivery of fabrics, leather, furs, or other materials to the clothing and footwear factories, the textile enterprise or tannery responsible for the delay will pay a fine of 0.4%, and after 20 days — 6% of the value of the raw and other materials that have not been delivered.

Clothing and footwear factories that are late in delivering the goods ordered by shops and wholesale and retail trade bodies will pay for each day's delay a fine of 0.3%, and after 20 days — 5% of the value of the clothing or shoes that have not been delivered.

Should the shops and trade bodies fail to take the goods they have ordered in the course of a month, they will pay the producer a forfeit of 5% of the value of the goods that have not been taken.

Thus, the transfer of light industry enterprises to the new system of planning and marketing output presupposes, on the one hand, a material interest on the part of the enterprises and their workers in the production of high-quality consumer goods, and, on the other hand, financial control over the results of production activity. Under these conditions they actually will work not for the shop, but for the consumer through the shop.

Pravda, January 22, 1965

Material Stimuli and Production

(Problems of the Economics of Agriculture)

L. Kassirov

Questions related to improving the management of industrial enterprises have been widely discussed in the press of late. These matters are just as important for agriculture. The main thing is to bring into action economic stimuli that will further the growth of output, satisfy more fully the population's needs in foodstuffs, and make it possible to channel the incomes of the collective and state farms both to meet the needs of the whole people and to ensure expanded reproduction in each farm. The solution of these problems depends on many factors. And these factors will be the subject of the present article.

The present system of planning needs further improvement. Its principal drawbacks are its cumbersomeness and the administrative nature of plan targets. The targets contain a multitude of indices which obstruct initiative on the part of management. Therefore, the proposal to establish — in addition to targets for the volume and structure of output — a unified index, which would serve as a criterion for appraising the work of any farm, is justified. Profit could well serve as such an index for agricultural enterprises. In addition, enhancement of economic methods of management requires improvement of the relationship between the state, on the one hand, and collective and state farms, on the other. This refers, above all, to the system of delivery prices.

The existing price system has a number of defects which lower its effectiveness as an economic means of stimulating production. One of its weak spots is the inadequately substantiated relationship of prices for various products. This produces the problem of "profitable" and "unprofitable" products. The average level of profitability assured by the existing

281

prices for the collective farms of the Russian Federation has varied in recent years from 3% to 100% and more, depending on the product. Certain products yield only losses.

A price policy that provides a different level of profitability for different products is not only permissible, but even advantageous. It should stimulate production in accordance with the planned rates of growth. However, in some cases the existing prices do not achieve this. For instance, the economic plan provides for higher rates of growth of milk output than of sunflower or sugar beet production. Yet the existing prices stimulate the output of these crops more than that of milk. In this case the price, instead of being a powerful stimulus, hampers the fulfillment of the plan targets. Obviously, the only way out under the circumstances is administrative regulation of the proportions of production.

What methods, other than administrative ones, can be used to compel the collective farms of Volgograd Region to fulfill the plan targets for the sale of potatoes if the cultivation of potatoes in that zone, as distinguished from other crops, only brings losses?

If a collective or state farm is given plan targets for the sale of a certain product, one should assume that the purchase price of that product assures the farm a profit. It is precisely profit that is the sole source of expanded reproduction. But prices should not be determined on the basis of this principle alone. This would put too great a burden on the state. What should be done? It would be possible, for example, to reduce the potato purchasing plan for Volgograd Region and increase it for Gorky or certain other regions, where the cost of growing potatoes is three or four times lower and the yields are much higher. The important thing is to determine which regions and zones can fully meet the state's needs in that product at the lowest cost.

Who is in the best position to determine all the possibilities and, above all, the reserves of production? The collective and state farms themselves are best able to do this. Therefore, it would be advisable to draw up state plans on the basis of proposals submitted by the farms. These plans, in summary form, will show the interest of the farms in growing one product or another and, when necessary, they will permit the adjustment of prices in keeping with state requirements. It follows, therefore, that the price system should not rule out the problem of the "profitable" or "unprofitable" product — far from it. The interests of the national economy should be considered first in establishing prices. This is the primary condition for the advancement of economic methods of managing the collective and state farms.

Increased state investments in agriculture have been accompanied by a substantial increase in output, as well as in the level of collective farm incomes and the earnings of the farmers. However, owing to the defects in the zonal price system, the income differences between farms of various zones are increasing instead of decreasing.

The prices of many products do not correspond to the differences in outlays resulting from the objective conditions of production. Here is a case in point. The cost of grain grown by the collective farms of Krasnodar Territory is nearly one-fourth of that in Kirov Region, where the purchase price is only 30% higher. As a result, the collective farms of Kuban have a high profitability (approximately 300%), whereas those of Kirov Region barely cover their costs.

Consequently, the possibilities of the collective farms in different zones as regards labor remuneration and expanded reproduction are far from equal. Thus, in the last five years (1959-1963) the collective farms of the Northern Caucasus and the Central Black Earth Zone obtained a per hectare net income that was nearly four times greater than that of the collective farms of the northwestern and East-Siberian regions of the country.

Unfortunately, certain personnel of the USSR State Planning Committee are opposed to any and all changes in the existing zonal price system. They believe that it is possible to equalize the economic conditions for collective farms of various zones by intensifying farming and rendering state aid to the lagging farms. Of course, intensification will raise the general level of production in all the zones and will gradually even out the distinctions among them. But intensification presupposes that funds will first be invested in those areas where they will yield the most rapid effect, where they will ensure higher incomes both for the state and the farm. Therefore, the problem of improving the zonal price system, far from being eliminated, becomes more important with the intensification of agriculture.

As for aid to the lagging farms, it will apparently be advisable to render it through cost accounting media, above all through the price system, so as not to encourage parasitical tendencies.

The question of profit distribution plays an important role in the operation of agricultural enterprises. It should be pointed out that the collective farms still fail to take account of this factor both in planning and in bookkeeping. In the state farms, part of the profit is used to expand production assets and to give the workers material incentives. However, the share of profits going for the expansion of production assets comprises an insignificant percentage of the capital investments. In 1963 the state farms of the Russian Federation covered from their own sources only about 3% of the total capital investments, the remainder being derived from the state budget.

This procedure contains an essential defect. Since the funds allocated to the state farms from the budget for the construction of farm buildings and the purchase of equipment are granted without charge, farm leaders have little interest in achieving economies in their use. It would seem proper, in our opinion, to shift from budget financing of capital investments and construction to the granting of long-term credits to state farms.

State farms are now given a so-called credit for the introduction of new equipment. The credit is reimbursed from the additional profits derived from the corresponding measures. To receive such a credit, a state farm must substantiate its request economically, indicating the purpose for which the sum is required. This compels the state farms to adopt a wholly responsible attitude toward the size of the sum requested.

However, the share of this credit in the overall sum of capital investments in state farms has been insignificant up to now. Of course, if a shift is made from gratis financing to the granting of credits for capital investment, the time required for recouping the credit will be greater. Incidentally, this is one of the conditions for the further improvement of prices. If capital investments are made at the expense of profits that are assured by prices, it is obvious that the need for budgetary allocations to the state farms will in large measure be eliminated. The state will then chiefly finance measures of a general character, such as land improvement projects, electrification, and the like.

This will free the budget of unnecessary expenditures. In addition, a certain portion of the funds now allocated to state farms under gratis financing could well be directed along another channel — through delivery prices — and thereby consolidate the economic foundations of cost accounting. Then the leaders of state farms will see an additional source of accumulation not in the budget funds of the state, but in the results of the operation of their own farms.

Profit still plays an insignificant role in creating material incentives for state farm workers. In 1963 the state farms of the Russian Federation used less than 2% of their total profits to award bonuses. But even so, it is not so much a matter of the smallness of the sum used for encouragement as of the system of forming the state farm's bonus fund.

The right to allocate means to the fund of the state farm is conditioned not only by the existence of profit, but also by the fulfillment of a number of other indices. It so happens that if the farm fails to fulfill any one of these indices, even though the indices on the whole are high, it is deprived of the right to allocate funds to encourage the workers materially. Thus, in the last few years not a single state farm of the Kaluga, Tula, and several other regions was able to assign any means to the given fund.

Only half of the allocations may go for bonuses to farm workers. The other half must be spent on expanding the fixed assets, although special allocations are made for this purpose.

Is this practice sound? We do not think so. Allocations to special funds should depend solely on profit. The greater the profit of the farm, the more it should be able to spend on its own needs. To give the workers an interest in the end results of production, one can hardly find a better index than profit. Profit reflects both the quantity and quality of the produce, its marketability and the cost of production. Moreover, the state farms must be given the absolute right to dispose of that part of the profit which is left to them.

Profit calculation is more necessary for the collective farms than for the state farms. The rate of growth of production and proper use of material incentives depend on the correct calculation and distribution of profit. If the collective farm does not know the sum of its net income (or, indeed, whether the farm is being operated at a profit or a loss), how can it determine where to allocate funds, and how much to allocate, after the costs of production have been covered? In the absence of profit calculations, some collective farms remunerate labor at the expense of weakening their production facilities, while other farms undertake construction beyond their means and, as a result, underpay the farmers.

It should be pointed out that calculation of profit in the collective farms is also essential for substantiating their contribution to the total income of the state. Collective farms are now taxed irrespective of their profit. This approach cannot, of course, give a sufficiently objective picture of the participation of collective farms in state expenditures. Net income should be the criterion for this.

By improving the economic methods of managing the collective and state farms, the state will be in a position to regulate material stimuli and to plan agricultural production better and more effectively.

Izvestia, May 28, 1965

The First Chord

V. Vukovich

Mine No. 9, "Velikomostovskaia," in the Lvov Economic District, reached its planned rate of output early last year. Its indices would cause it to be classified as an average enterprise, of which one usually hears less than about a leading or lagging unit. But suddenly people began to talk about Mine No. 9 at the economic council and the regional committee of the party in Lvov, and at the Institute of Economics, the Ukrainian Economic Council, and the State Planning Committee of that republic in Kiev.

If you have ever had frank talks with planners and production personnel about the relations between them, you know how many complaints they have about each other. Virtually every person in production holds that all planners do is to raise production targets. The planners, on the other hand, hold that operational personnel want an easy life.

These were approximately the relationships that used to exist at "Velikomostovskaia," No. 9. But today these mutual recriminations have ceased. The factors giving rise to them have disappeared. They are gone, thanks to an economic experiment which the miners undertook last October.

The sense of the experiment was to verify in practice the possibility of enlarging the economic independence of an enterprise while simultaneously increasing the material incentive of each individual to improve economic and technical indices.

December was the month of strategic preparation. The miners were aided by experts from the combine and the economic council and by scholars of the Institute of Economics of the Ukrainian Academy of Sciences. The economists decided to evaluate the work of the miners with respect to fulfillment of the output plan, conventional profitability (reduction of the loss at which the operation is conducted), and the quality of the coal. It was also necessary to provide for something else.

The fact is that the mine at which the experiment was begun is an island in an ocean in which the previous relations among enterprises continue to operate. This factor was also considered.

Five months have passed. During this period the mine has jumped forward to a degree that even the most fervent advocates of the experiment had not expected.

"It's best if we start to discuss the first steps," the mine manager, Iu. S. Taraskin, begins, "with that which was most important, the change in planning practices. Previously the trust had established more than thirty indices for Mine No. 9, as for all other mines. It was not unusual for these indices simply to be incompatible with each other. Moreover, the indices would then be changed. Last year, the planned production cost was changed seventeen times, and the total wage fund and number of workers — thirteen times. It's hard to maneuver under such conditions!

"The experiment compelled a different approach. In January, 'Velikomostovskaia,' No. 9, was given only three norms, fixed for the year: the output plan by quarters, the state subsidy per ton of coal sold, and the ash content standard. The trust also confirmed a plan for the extension of mining operations."

"All other indices," Taraskin continues, "we developed ourselves, with maximum use of our resources as the goal. A firm foundation for the experiment was provided by the new bonus system, the aim of which is to give each employee and the staff as a whole a material interest in obtaining the best possible results.

"The experiment offered scope for initiative and for rational economic maneuvering. It sharpened the desire to work to the utmost.

"The mechanism by which people establish strenuous plan goals for themselves is not complicated. Norms are applied for each drift based on the findings of time studies, with consideration for geological and mining conditions, and for the technology employed. These norms must be rigidly adhered to. However, each man knows that if he mines more coal than the norm, a differential bonus system will come into effect. Thus, the personnel have an interest in undertaking a plan that exceeds the norm. The bonus is increased progressively in this case. Generally speaking, the more the norm has been exceeded, the higher the earnings.

"It is not only the workers, but the executives, engineers, and technical personnel of all services and departments of the mine who have an interest in strenuous plan targets. Each of them knows that if the plan goal for his sphere of the operation is increased over the norm, the bonus will be larger. This is the system that ended the recriminations between the planners and the operational personnel. The coal producers themselves try to go over the norms, collectively improve work methods, and constantly seek new untapped reserves, because this is profitable to each of them, to the mine as a whole, and to the state.

"Independence in planning and the levers of material incentive have made themselves felt in results. 'Velikomostovskaia,' No. 9, has become the leading mine in the trust. Instead of the planned output of 1,520 tons every twenty-four hours (which is higher than the designed capacity of the mine), it yielded 2,041 tons a day during the first quarter. Labor productivity rose 35% over last year's figure. There is no longer anyone who fails to meet the norm. Wages have also gone up considerably. It is very important, in this connection, that wages per ton of coal have dropped by three kopecks."

"The force of this experiment," comments L. S. Striukovskii, Chief Economist of the mine, "lies also in the fact that the miners are learning to calculate everything, to calculate how to produce more coal with lower expenditures. How much used to be said at meetings and in orders to the effect that metal props, clamps, rails, accessories, bolts, and pins should be reused! But at the mine face people reasoned: 'Why should we bother with old stuff; let's order new things.' Now, before leaving a mined-out stope, they remove everything in it, repair it, stack it carefully, and invariably make use of it.

" 'It's not only tons we have to figure,' people say now at the faces, 'but what they cost us.' "

All the cutting-and-loading machines used in this mine are considered old because they have undergone major overhaul. But how they operate! Each unit produces, on the average, 6,879 tons per month, or 929 more than the average for the trust as a whole.

During the first quarter the mine saved — without loss to production technology or safety — 800 cubic meters of grade-one timber. The production costs for a ton of coal in the mine are 10% lower than last year, and the state subsidy was reduced by 130,000 rubles during the first quarter.

Where did the net profit go, and what was it used for? Half went to pay bonuses to the personnel, and the rest was divided into two equal parts: one part went to the state budget, and the other to the enterprise fund. The money in the latter goes for the cultural, everyday, and other needs of the personnel.

Changes, changes! They are many at "Velikomostovskaia," No. 9. A great deal is now being said about the experiment and the first results it has yielded. The belief is that it must be continued so as to obtain stable results and, in addition, to find an answer to the question as to how to improve the new system.

Guests from the Donbas, Vorkuta, and other mining districts come to this mine, which has become a pioneer in economic exploration, to study its experience.

Pravda, June 1, 1965

A Stimulus to High Quality

(Instructive Economic Experiment at Sverdlovsk Plants)

V. Bliukher and N. Il'inskii

The strength of our economic system lies in the unity of interests of the state and the enterprise, of the entire collective and each working person. However, economic conditions still exist that restrain the initiative of staffs in the struggle to improve the quality of output.

The stable prices and fixed targets for reducing costs of production do not, as a rule, make allowance for the extra expenditures involved in improving the quality of output. But improved products are not born of themselves. That is why the manufacture of new and improved goods is often unprofitable for a factory.

We have analyzed the profitability of 52 machine-building enterprises in the Urals. It turns out that the higher the ratio of new products to total volume, the lower the rate of profit. The profitability at enterprises with less than 10% of their output accounted for by new products is 27%. However, at enterprises where new items are half the total output, profitability drops to 11%.

Let us assume that a machine-building plant has improved the quality of its serial production by its own initiative. What are the economic consequences for it? Let us say, frankly, that they are not reassuring. Additional expenditures of labor and monetary resources occur and, consequently, the costs of production become somewhat greater than the plan had envisaged. This in turn results in the loss of bonuses, failure to meet the profit plan, etc.

The authors are, respectively, Chief of the Machine-Building Administration of the Mid-Urals Economic Council and a Pravda correspondent.

291

In order to provide enterprises with a real interest in improving the quality of their products, a number of urgent steps must be taken. The most important would be to incorporate indices in the production plans of enterprises and economic councils that will take full account of the productivity, reliability, service life, and operating costs of the machines manufactured. But the question arises of what to do if no technique exists for evaluating the quality of most types of products. There is one way out: to seek new solutions in the field of economics, to develop quality indices.

This has impelled scientists and machine builders to conduct an experiment at one plant in planning economic indices with due consideration for the quality of the products manufactured. The Sverdlovsk Ball Bearing Plant was chosen. The workers, engineers, technicians, and scientists decided to build long life into the ball bearings and thus to improve sharply the dependability and service life of any machine. Also a factor was the consideration that evaluation of the quality of this product is comparatively simple, and that it can be measured by a single basic index — service life, i.e., the number of hours it actually functions.

In March of last year the Mid-Urals Economic Council adopted the decision to conduct this experiment and made additional personnel and money available for the purpose. The second quarter became the period of preparation for the new method of work. A new processing technology for bearings parts was developed to yield higher precision, and nine new operations were added which increased the labor requirements per unit of output by 33%. The Sverdlovsk Machine-Building Research and Development Institute designed, and the plant built, special automatic machines to polish roller surfaces. As a consequence, surface finish was upped two grades.

Scholars at the Department of Machine-Building Economics of the Urals Polytechnic Institute made their contribution to preparations for the experiment. Under the new experimental planning system, volume of production is determined not only by the usual indices, but also as the product of the number of bearings manufactured times their average service life (hours in operation), which is established at the testing stand. An index called "specific costs of production," i.e., production costs (planned or actual) per hour of bearing operation, was introduced to calculate product cost.

We are now in a position to cite the preliminary results

of this economic experiment. By the beginning of the second half of 1964, when the plant personnel set to work to achieve a major improvement in the quality of the bearings, a bearing cost 3 rubles 86 kopecks, and service life under intensified operating conditions on the testing stand averaged 257 hours. Thus, the specific cost of production (per hour of bearing operation) was 386:257 = 1.5 kopecks.

The change in production technology required additional expenditures. This led to a certain increase in the cost of producing the articles. Thus, in the fourth quarter of the past year, a bearing of increased service life cost the plant 4 rubles 12 kopecks. However, the entirely justified expenditures for introducing the additional technological operations that increased the efficiency of the products were recouped many times over. The service life of the most popular bearing has already reached 680 hours on the testing stand. This is more than twice the usual service life. Consequently, specific production costs per bearing have dropped to 0.6 kopecks. Considering the number of products whose service lives are increased as a consequence, the national economy saves more than a million rubles per year.

The experiment resulted in a fundamental change in the attitude of workers and specialists to matters of production quality. The Mid-Urals Economic Council has worked out a regulation on bonus payments to the personnel of the plant for reducing the specific production costs of bearings. As before, the source of this incentive money is the bonus fund for placing new equipment into operation. But it is being employed in a new way. The plant considers not the individual measure resulting in an increase in quality, but the results of the entire complex of measures.

It is interesting to note that the employment of specific production costs causes a significant change in the results of the economic functioning of the enterprise. During the fourth quarter of last year, the plant exceeded its output plan for bearings by 4.6%. However, if we consider the increased service life of the bearings, the plan assignment for total service life was exceeded by 13.9%.

The same holds for the plan for reducing costs of production. Without taking into account output quality, the cost of production was reduced by 1.9%, but with allowance for increased service life the figure was 13.9%. Consumers are willing to pay for the additional outlays made to produce bearings with a longer service life, and sales prices have been increased somewhat. As a consequence, the bearings with the longest service life have become the most profitable for the plant to manufacture. They also benefit those who manufacture and utilize machinery.

We hold that the first results of the economic experiment permit far-reaching conclusions to be drawn. The most important is that, when favorable economic conditions are created, the initiative of enterprise staffs increases in the struggle to improve quality of production. That which is advantageous to the state becomes advantageous to the enterprise. The "specific cost of production" index, proposed for planning and evaluating the functioning of an enterprise, has justified itself to the full and may be applied to other machine-building enterprises if allowance is made for the distinctive characteristics of the products. The index serves as a stimulus for high-quality output and assures that the interests of the enterprise and the national economy are in harmony.

On the basis of the experience thus acquired, the personnel of the Sverdlovsk Tool Works has also undertaken an economic experiment. The basic index of a tool's quality is durability, i.e., a time factor, the total service life until it is entirely worn out. The plant's engineering staff and the Durability and Dependability Laboratory of the Sverdlovsk Machine-Building Research and Development Institute have worked out an effective method of determining tool durability. The efforts of the toolmakers have yielded results: increased durability and lower specific costs of production (for one minute of service life) make possible an annual saving of 252,000 rubles. However, the additional expenditures come to only 18,000 rubles.

The personnel of the Alapaevsk Machine-Tool Works and the Department of Machine-Building Economics of the Urals Polytechnical Institute have developed an original technique for evaluating machine-tool quality, a testing procedure, and economic indices. They have found an overall index, the "quality index." It is determined by a formula which includes relative change in indices of productivity, service life, and dependability of products, with allowance for the economic yield consequent upon the introduction of additional technological operations. In this case, the specific cost of production is determined by dividing expenditures by the quality index.

It is clear that economic calculations require the most careful verification. How is this to be done? The Mid-Urals

Economic Council possesses no surplus resources for purposes of experimentation. The Sverdlovsk machine-building industry has therefore turned for assistance to the RSFSR Economic Council. But a year has passed and the matter has yet to be resolved.

We believe it desirable to expand the experiment in utilizing the specific cost of production and, on the basis of study of its results, to introduce this index into planning practice as of 1966. The same should be done with the other results of economic research aimed at improving the quality of output. The next move is up to the USSR Economic Council and the USSR State Planning Committee.

It seems to us that the USSR State Planning Committee, the state committees for various fields of machine building, and their leading research institutes would be doing the right thing if, in the briefest possible time, they were to develop quantitative indices for production quality and a technique of evaluating them for all branches of machine building.

Komsomol'skaia pravda, June 3, 1965

The Wings of an Experiment

L. Pekarskii and S. Anufrienko

The Maiak Clothing Association in Gorky and the Bol'shevichka Clothing Association in Moscow have been functioning for nearly a year on the basis of direct contractual ties with trade organizations. A good deal, almost entirely positive, has been written and said about this experiment. Today everyone admits that direct contractual ties between the supplying enterprise and the store definitely constitute something that offers good prospects and is economically beneficial.

The results of the experiment may be "felt" by every resident of Moscow and Gorky. All one has to do in order to confirm this is go to a large clothing store. The products of these associations are no longer lying around on the shelves.

This progressive method of operation is becoming ever more widespread. Starting in the third quarter of this year, all enterprises in the clothing and footwear industry of Moscow and the other largest cities of the country will be converted to producing their ready-to-wear products in accordance with orders received from stores and wholesale and retail trade organizations.

In order to meet the needs of clothing and footwear enterprises for raw and other materials more effectively, a number of textile mills and tanneries are converting to operation on the basis of orders received from the enterprises.

Thus, the experiment has already reached the highroad of mass-scale application. And so the time has now come to see whether we are getting a full return from direct ties between industrial enterprises and trade organizations.

The authors are associated, respectively, with the Scientific Research Institute of the USSR State Planning Committee and the Institute of Economics of the Academy of Sciences of the USSR.

It must first be remarked that even today many economic councils and trade organizations continue to impose, from
above, lists of products to be manufactured upon enterprises
working on the direct contact system and are incapable of
ridding themselves of the habit of allocating commodities by
edict. But the very sense of the new system of planning is that
stores and factories bear the responsibility themselves for what
they produce and sell. However, this condition is often violated.

By way of example, the Clothing Industry Administration
of the Volga-Viatka Economic Council ordered the Maiak firm
to manufacture, during the second half of 1964, a considerably
larger number of leatherette coats than the stores had ordered.
Nor has the Clothing Industry Administration of the Moscow
City Economic Council let the reins of management by edict
out of its hands. Contrary to the procedure that has been
established, it continues to impose production quotas for
grades of goods upon the Bol'shevichka factory.

The trade organizations are making their contributions.
The largest stores in the capital — GUM, TsUM, Sintetika, and
Moskva — wished to order more than 5,000,000 rubles worth
of products from Clothing Factory No. 52 during the second
half of 1965. The factory was willing. But the trade agencies
followed the beaten track: the stores were assigned only half
the amount they had ordered.

The management agencies seek to justify their infringe-
ments of the rights of enterprises working under the new
method by alleging that this is necessary if the plan for the
economic council as a whole is to be carried out. Such arguments are untenable. What the consumer wants is not a barrage
of figures, but attractive, comfortable footwear or clothing.
That is why production was made dependent upon the orders
from the trade network, and why the regulation on direct
contractual ties stated that the economic councils are not to
set quantity and quality targets for the enterprises.

There is also another factor making for diminution of the
effectiveness of operations conducted in accordance with direct
ties. The chain of enterprises working under direct contracts
is incomplete. Machine-building and chemical plants are not
included. But it is precisely because of the lack of quality
chemicals, vivid and fast dyes, dependable fasteners and needed
equipment that textile mills are often unable to fill the orders
of the clothing industry.

The largest cotton textile combine in the country, Moscow's

Trekhgornaia manufaktura, sometimes lacks the dyes it needs. To avoid putting out fabrics of the same range of colors month after month, the combine's management has to engage in all sorts of "cooperation" — to exchange dyes with other mills, etc. But this, of course, is no solution to the problem. The situation would change fundamentally if the textile enterprises were to establish direct contacts with chemical plants.

Textile enterprises are often supplied equipment that is low in productivity and is unreliable. Thus, the output of the Ivanovo Textile Machinery Plant — one of the principal producers of finishing equipment — is much inferior to that of plants abroad. Yet it is slow in putting new models of this equipment into production.

The clothing workers are in no better position. Machines for sewing clothing from the latest fabrics and synthetics are not available. It is said that there is no one to make them. What do they mean, no one! The Podol'sk Machinery Works manufactures large numbers of home sewing machines, although the stores are utterly overstocked with them. This enterprise could undertake the manufacture of the new equipment. Here is a situation in which responsibility for output on a cost-accounting basis, such as would develop under the direct-contact system, would help.

Factories working on the direct-contact system have been granted considerable independence in planning. It is therefore very important to find the most effective method of encouraging good results in production. But what is happening today? At Maiak and Bol'shevichka, bonuses to the leading engineering and technical personnel and to office personnel are paid as follows. Those working in the management of the association are paid bonuses for fulfillment and overfulfillment of the profitability plan, while personnel in the shops get them for fulfilling the output plan and for above-plan improvement of the quality of the product.

It goes without saying that such a procedure for material incentive can hardly be regarded as the best possible. For the smaller the plan assignment, the easier it is to fulfill it and obtain a bonus. This means that in concluding contracts with stores, the enterprises have an interest in understanding their capacities and concealing reserves for expanding production.

The situation is even worse with respect to another form of reward. Reference is to the enterprise fund, i.e., to the means which are used to pay bonuses and to improve cultural, housing, and other living conditions. Four percent of the profit earned by the association is paid into this fund. But this 4% adds up to one sum when costly products are manufactured, and to a considerably smaller sum when inexpensive items are produced. Therefore, it benefits a factory, for example, to make more expensive suits. But the need is that an association have an identical interest in producing whatever goods are demanded by the population.

Of course, "differential benefits" can be done away with by an improvement in the system of prices. However, in the first place, that is a separate problem which is not going to be solved soon, and enterprises are switching over to direct contacts right now. In the second place, differences in the profitability of products will remain in any case. They will arise out of the introduction of new materials, the mastery of new output, and so forth. Moreover, children's clothing will always be cheaper and less profitable than adult wear. Therefore, the system of incentives must be improved so that it takes fuller account of the real services performed by the staff.

Direct contacts will be most effective when the study of current and long-range demand is placed on a scientific basis. But demand must not only be studied. It is also necessary to educate consumer taste with the aid of advertising.

Unfortunately, this "engine of trade" is still operating at a small number of "revolutions." The number of commodities on the counters is constantly increasing, and the consumer finds it harder and harder to choose among this variety. For example, during this past year, Moscow enterprises alone produced 3,156 new models of clothing and hats, and 500 new models of footwear. How did advertising respond to this? It didn't. The consumer was left to face the array on the counter unaided.

Yet advertising, by influencing the taste of purchasers, is capable of easing the planning of production and the study of consumer demand. It can help us to avoid random interest in one item or another. Such interest results in frequent changes in orders and higher production costs. Proper advertising accelerates commodity turnover. This is also very important. Every additional day that commodities are on the shelf causes enormous losses to the country.

We all have an interest in good advertising. But improvement of its artistic and technical level will require an increase in the money spent on it, which is now 0.02% of the total value of retail trade. It is hardly necessary to prove that such expenditures will pay for themselves with interest.

Direct contractual ties between manufacturers and stores are a recent development. But they have been developing at a heroic pace. And if we continue to regard this law-governed process as a "newborn baby," we may find ourselves facing the fact that the child has grown up, but has nothing to wear.

Trud, July 8, 1965

To Develop Direct Production Contacts

The progress of science and technology has opened new opportunities for improving the quality and variety of output. Improvement in quality has become, under present circumstances, one of the important and principal tasks of the economy. And it must be borne in mind that we not only need improvement in the quality of the products we are putting out, but systematic and, what is most important, rapid renovation of products in accordance with rising demand.

That is why it is essential that each enterprise consider the growing demands made upon its products from the standpoint of quality and reliability, and also in terms of new technical possibilities. In order for factories and mills to get a better sense of market conditions and changes in consumer demand, it is necessary to engage, on a large scale, in the establishment of direct contacts among enterprises, associations of enterprises, and the trade organizations that sell their products.

The first experience in organizing such contacts between suppliers and consumers has already been gathered.

The economic experiment conducted by the Bol'shevichka Clothing Association of Moscow and the Maiak Clothing Association of Gorky demonstrates persuasively that the new procedure for the planning and marketing of output represents a correct path that should be followed by many. A USSR-wide Tuesday conference conducted by Trud, which was described in detail in yesterday's paper, was devoted to analysis of this valuable experience and to spreading it.

With the conversion to direct contacts, under which both the schedules for clothing manufacture and the quantity and quality indices for the products made came to be determined not "from above," as had previously been the case, but exclusively in accordance with the orders of the trade organizations, the manufacturing enterprises gained a degree of independence and felt more responsible. There began a reorganization of the entire process of production so as to protect the trade network against a piling up of unsold goods. The matter was posed as follows: provide the population with attractive, stylish, good, and inexpensive clothing, and make sure that production is sold in its entirety and does not accumulate in the warehouses.

Thus, the consumer himself began to determine the line of clothing, styles, and sizes, through the medium of the trade network. And the stores began to carry what people were looking for and wished to buy. Today the stores in Gorky are ordering twice as many models of clothing from the Maiak firm as before, and the assortment has increased considerably, but coats, dresses, and suits no longer lie around on the shelves.

The newly founded experimental department of the firm has become what is in effect a laboratory for the scientific organization of work at the enterprise. The work of designers and sample tailors has been reorganized, planning within the department has been changed, work quotas have been revised, the production process has been refined, and additional conveniences have been provided for the women workers. This was quickly reflected in increased quality and expansion of the assortment of products. Today the clothing workers shift quickly and flexibly from one model to another, and, what is not unimportant, do not lag behind style.

Direct production contacts demonstrate that this new method of management enhances production efficiency, reduces spoilage, and makes labor more effective. Whereas, previously, unsold goods in the warehouses and stores of Gorky considerably exceeded the permissible limit, today they come to a total value considerably below that allowed for in the plan.

Economic stimuli are transforming the nature of production in clothing firms before our very eyes; they are changing people's attitudes toward the work assigned to them and are eliminating previous concepts as to the "advantages" and "disadvantages" of plans, orders, and assortments of goods. Businesslike qualities, cost accounting, and initiative are acquiring rights of citizenship. And herein we find one of the notable features of the new economic relationships between industry and trade.

The question is now being discussed of introducing this progressive form of management — direct contacts between suppliers and consumers — into all branches of production, including those producing machines, equipment, and metals.

The importance of this has been persuasively described

in the pages of Trud by worker-correspondents of the Elgava Machine-Building Plant. The Latvian Economic Council ordered this enterprise to manufacture 2,500 tractor axles although this work is utterly unrelated to the business of the plant, and each axle would cost approximately sixteen rubles more than the sales price. The economic council insisted upon the production of these "golden" axles while hundreds of enterprises throughout the country waited for the plant's proper products — automatic lubricating equipment and brakes.

On their own responsibility and risk, the Elgava people subcontracted the axle job to an enterprise producing automotive and tractor parts, which manufactured the axles for them willingly, without difficulty, and at a low price. It is not hard to imagine what the sort of "planning" and "management" engaged in by the economic council would lead to.

The warning issued by the Trud worker-correspondents was heeded in the USSR Economic Council. A plan has now been developed for the long-term attachment of major consumer-enterprises to supplier-enterprises, including the Elgava Machine-Building Plant. Enterprises are now gaining the opportunity to work out contracts directly with their consumers for the delivery of the bulk of their products, bypassing the stage of receiving commodity allocation schedules [raznariadki] from the supply and marketing agencies of union republics, economic councils, economic districts, ministries, and departments.

Life has convincingly demonstrated that the time has come to draft production plans on the basis of orders from consumers, with provision for the establishment of direct connections between industrial enterprises and trade organizations. The transition to this form of planning will doubtless be a step forward, as planning based upon orders will be more realistic, concrete, and closely tied to the needs of the economy and the population. This has been confirmed and demonstrated most clearly by the USSR-wide Trud Tuesday conference in Gorky.

Life has also demonstrated that it is necessary to follow the path of planning in accordance with the orders of consumers not only in industry manufacturing consumer goods, but in other branches of the economy as well.

New forms of relationships between enterprises must be introduced into the economy more boldly and decisively. It is necessary to clear the way more rapidly for everything that is progressive, effective, and economical. Direct production contacts are the proper course. If we follow it, we will move toward success with even greater confidence.

Soviet Life, July 1965

Are We Flirting with Capitalism?
Profits and "Profits"*

E. G. Liberman

In its February 12 issue this year Time magazine carried my picture on its cover, with the prominent caption, "The Communist Flirtation with Profits." The cover story, entitled "Borrowing from the Capitalists," made many references to my writings and statements, and drew conclusions vastly different from those I make. I therefore asked the editors of Soviet Life and Ekonomicheskaya Gazeta to permit me to comment on the Time article in their publications. To do a proper job, I shall have to go rather deeply into the essential character of profits.

***

Profits are the monetary form of the surplus product, that is, the product which working people produce over and above their personal needs. The surplus product is, therefore, an expression of the productivity of labor. Primitive man ate or used up what he produced. As civilization and technology progressed, labor began to create not only the equivalent of the working people's means of subsistence but something more. This something more was the surplus product, the very same surplus product that supports the entire nonproductive sphere, from the watchman to the banker and cabinet minister.

But the surplus product is also the source of means essential for the development of society. That applied to feudal and capitalist society, and it applies to socialism and communism.

Under socialism products and services are also produced

---

*Reprinted by permission of Soviet Life. Translation by Soviet Life, Washington, D.C.

as commodities and also sold chiefly for money. Therefore, the surplus product inevitably assumes the monetary form of profits. But since profits in our country are used in the interests of society, they become less and less an expression of surplus (unpaid) labor and come more and more to express socially necessary labor.

What is the difference between capitalist and socialist profits?

Bonus "for Risks"

The difference is not, of course, that private enterprise stands for profit while socialism "denounces" it, as economists in the West often claim. To make the difference clear, let us examine 1) how profit is formed, 2) what it signifies, and 3) for what purposes it is spent.

From the private entrepreneur's viewpoint, all profit belongs to the capitalist. To support this view, economists built the theory of the three factors that create value: capital, land and labor. In The Theory of Economic Development Joseph A. Schumpeter says that profit is everything above cost. But this "cost" includes "wages" for the labor of the entrepreneur, land rent, interest on capital, as well as a bonus "for risks." On top of that, the entrepreneur reaps a profit if he succeeds, by a new combination of production elements, in reducing the cost to below the existing price level.

What kind of "combination of elements" this is can be seen from the fact that the main part of the profit under the private enterprise system now comes not so much from production as from the process of exchange. For instance, high profits come most readily from advantageous buying of raw materials, the raising of retail prices, the tendency of unemployment to lower wages, nonequivalent exchange with developing countries, the export of capital to countries where wages are low, the system of preferential tariffs and customs duties, raising the prices of stocks on the stock exchange, and so-called Grunder (speculator's) profits.

Does Money Smell in the Soviet Union?

All those sources of profit are ruled out in the Soviet Union owing to the very nature of socialism, under which there is neither private ownership of the means of production nor stock capital and, consequently, no stock market. The level of payment for labor depends on its productivity and is regulated by law. The prices of raw and other materials are planned; market conditions that could be taken advantage of in purchasing raw materials or hiring labor do not exist. Nor can the prices of finished articles be raised by taking advantage of market conditions. Exchange with other countries is conducted on the basis of equality and long-term agreements.

Legend has it that the Roman Emperor Vespasian decided to impose a tax on public toilets when he saw that his treasury was running low. His son Titus, who later succeeded to the throne, waxed indignant at such an evil-smelling source of revenue. Vespasian then held up to his son's nose the first receipts from the toilet tax. "Non olet!" ("It doesn't smell!") Titus exclaimed in surprise. Ever since then the view that "money doesn't smell" has been gospel in the commodity world. Indeed, under private enterprise nobody really cares how money is made. The important thing is to make it, the important thing is how much of it you can make.

But in the Soviet Union "money does smell." That will be seen if we look into the nature of profit. In our country profit testifies, in principle, only to the level of production efficiency. Profit is the difference between the selling price of articles and their cost. But since our prices, in principle, express the norms of expenditure of socially necessary labor, the difference is an indicator of the comparative economy with which an item is produced. Behind Soviet profits there is nothing except hours of working time, tons of raw and other materials and fuel, and kilowatt-hours of electrical energy that have been saved. Our profits cannot "smell" of anything but that. We do not justify profits obtained through accidental circumstances — for example, excessive prices — and we do not consider such profit a credit to the factory or other enterprise which makes it. We look on such profits, rather, as the result of an insufficiently flexible practice of price fixing. All such profits go into the state budget, without any bonus to the enterprise concerned.

Capitalist profit is a different matter altogether. As the reader knows very well, profits in the West can indicate anything under the sun over and above purely technical and organized efficiency. Commercial dexterity, successful advertising, profitable orders for military production — that is what the history of present-day big capital testifies to sooner than to anything else. Surely it must be clear that in essence and origin profit under socialism bears only a superficial resemblance to profit under private enterprise, while by its nature and by the factors to which it testifies it is fundamentally different from capitalist profit.

Where do profits go in the Soviet Union? First of all, neither a single individual nor a single enterprise can appropriate profits. Profits are not arbitrarily invested by any persons or groups for the sake of private income.

Profits belong to those to whom the means of production belong, that is, to all the citizens, to society. Profits go, first and foremost, for the planned expansion and improvement of production and scientific research, and to provide free social services for the people: education, health, pensions, scholarships. Part is spent on the management apparatus and, unfortunately, a rather large part goes for defense needs. We would gladly give up this last expenditure if a program of general disarmament were adopted.

Indicators in Industry

There is nothing new in that use of profits in the Soviet Union. Our enterprises have been making profits in money form for more than 40 years, ever since 1921. It is with these profits that we have built up our giant industrial potential, thanks to which we have moved to a leading position in world science and technology. And we have accomplished this without major long-term credits from other countries.

Why has the question of profits been so widely discussed in the Soviet Union lately? Not because profits did not exist before and are only now being introduced. The reason is that profit was not, and still is not, used as the major overall indicator of the efficient operation of our enterprises. Besides profit, we have been using a fairly large number of obligatory indicators — among others, gross output, assortment, lower costs, number of employees, size of payroll, output per employee, and average wages. The multiplicity of indicators hamstrung the initiative of the enterprises. Their main concern often was to turn out as great a volume of goods as possible since they would be rated chiefly on gross output. Furthermore, enterprises did not pay much attention to how they used their assets. Trying to meet their output quotas in the easiest way for themselves, they asked for, and received free from the state, a great deal of plant, which they did not always use efficiently or to full capacity.

How do we explain that?

Virtue Becomes Vice

For a long time the Soviet Union was the only socialist country. We stood alone, surrounded by a world in which there were many who wanted to change our social system by force. We had to build up our own industries and secure our defenses at all costs and in the shortest possible time. Such considerations as the quality and appearance of goods, or even their cost, did not count. This policy completely justified itself. The Soviet Union not only held its own in the war of 1941-1945 but played the decisive role in saving the world from fascism. That was worth any price. And that was our "profit" then.

But, as Lenin often said, our virtues, if exaggerated, can turn into vices. And that is what happened when we held to the same administrative methods of economic management after we entered the stage of peaceful economic competition with the industrial countries of the West.

We want to give every citizen, not only the well-to-do, a high standard of living, in the intellectual as well as in the material sense. In other words, we want everyone to have the fullest opportunity to develop his mental and physical capacities and his individual (I emphasize, individual, and not group) inclinations and interests. We want every person in our country to be able to do the work he wants most to do. We want to reach the point where it will not be possible to draw a hard and fast line between a person's vocation and his avocation.

Before we can bring people's intellectual capacities to full flower, we must satisfy their material needs, place goods and services of high quality within everyone's reach. These needs must be satisfied, moreover, with the lowest possible production outlays and the fullest possible utilization of all assets.

What Is the Way Out?

All that cannot be done through the old methods of administrative direction and highly centralized management. We must change over to a system whereby the enterprises themselves have a material incentive to provide the best possible service to the consumer. It is clear that to do this we must free the enterprises from the excessive number of obligatory indicators. In my opinion, the criteria for rating the work of enterprises should be: first, how well they carry out their plans of deliveries (in actual products); and, if these plans are fulfilled, then second, their level of profitability. I believe that out of their profits, enterprises should have to pay into the state budget a certain percentage of the value of their assets as "payment for use of plant." The purpose would be to spur enterprises to make the most productive use of their assets. Part of the remaining share of the profits would go into incentive pay system funds, the amount depending on the level of profitability. The rest of the profits would accrue to the state budget to finance the expansion of production and to satisfy the welfare needs of the population.

Plan, Profit and Bonus

Why do I choose profit as the indicator?

Because profit generalizes all aspects of operation, including quality of output. The prices of better articles have to be correspondingly higher than those of articles that are outmoded and not properly suited to their purpose. It is important to note, however, that profit in this case is neither the sole nor the chief aim of production. We are interested above all in products with which to meet the needs of the people and of industry. Profit is used merely as the main generalizing and stimulating indicator of efficiency, as a device for rating the operation of enterprises.

Yet Western press comments on my writings blare away about the term "profit," very often ignoring the fact that the title of my Pravda article of September 9, 1962, was "The Plan, Profits and Bonuses." They make a lot of noise about profit but say nothing about planning.

Actually, my point is to encourage enterprises, by means of bonuses from profits, to draw up good plans, that is, plans which are advantageous both to themselves and to society. And not only to draw them up but carry them out, with encouragement from profits. It is not a question of relaxing (or rejecting) planning but, on the contrary, of improving it by drawing the enterprises themselves, first and foremost, into the planning process, for the enterprises always know their real potentialities best and should study and know the demands of their customers.

The contractual relations with consumers or customers that we are now starting to introduce in several branches of light industry by no means signify that we are going over to regulation by the market. We have better ways of predicting consumer demand because we know the wage fund of the urban population and the incomes of the collective farmers. Therefore, we can draw up scientific patterns of the population's income and expenditure. In our country consumer demand, in terms of total volume, is a factor that lends itself to planning. However, the various elements of that volume — for instance, the colors of sweaters or the styles of suits factories should produce, or how best to organize their production — need not be the prerogatives of centralized planning but matters on which the stores and the factories concerned come to terms. Thus, the consideration of consumer demand and the planning of production are not only compatible in the Soviet Union but can strongly substantiate and supplement each other.

The Substance of Our Debates

The _Time_ cover story is full of contradictions. It admits that the Soviet people now have more money and that there is a growing demand for better and more fashionable clothing and for private cars. One would think that pointed to an improving economy, yet the article claims that the switch to profits is a result of "unsettling prospects," of "waste, mismanagement, inefficiency and planning gone berserk," and so on and so forth.

There are, of course, no few instances of waste and mismanagement in the Soviet economy, just as there are in private enterprise; think of the thousands of firms that go bankrupt every year. But in the Soviet Union we focus public attention on instances of waste and mismanagement. We publicize and criticize them openly. Some Western commentators take advantage of that fact. What better way can there be of distorting an over-all picture than to pick haphazard details and offer them as representative of the whole? The over-all picture shows that the Soviet Union increased its output by 7.1 per cent in 1964. _Time_ admits that this is a very good growth rate for a highly developed economy. It is not good enough for us, however. We are used to growth rates expressed in two digits. _Time_ does not mention that the reason for this 7.1 per cent growth rate, a relatively modest one for us, was the 1963 crop failure.

We are turning to profits not because we need a "sheet anchor." We are not in any danger. The fact remains, however, that we have to improve our methods of economic management. This is the substance of our debates and our searches.

The Main Function of Profits

Under socialism, profits can be a yardstick of production efficiency to a far greater degree than in the West, for in the Soviet Union profits follow, in principle, only from technological and organizational improvement. This also means that profits here will play an important but subsidiary role, like money in general, not the main role. After providing a yardstick of production achievement and a means of encouraging such achievement, profits in the Soviet Union are used wholly for the needs of society. They are returned to the population in the form of social services and expanded production, which guarantees full employment and better and easier working conditions for everyone.

In the Soviet Union nobody accumulates profits in money form — neither the state nor enterprises. This is an important point to grasp. If, for instance, at the end of the year the state as a whole has a surplus of budget revenue over expenditure, the surplus does not stay in the form of accumulated currency but is immediately used for two purposes: 1) to increase State Bank credits for material stocks, in other words, the surplus takes the material form of expanding inventories in production or trade, while money only measures this increase, and 2) to withdraw paper money from circulation, that is, to increase the purchasing power of the ruble on the free collective farm markets, where prices are determined by supply and demand.

Consequently, profits cannot become either capital or hoarded treasure in the Soviet Union. They are not, therefore, a social goal or a motive force in production as a whole. The motive force in production under socialism is the satisfaction of the steadily growing material and cultural needs of the population. However, profit can be, and should be, an indicator (and the key indicator, moreover) of production efficiency. It should serve to encourage workers to raise their efficiency. But it should be understood that encouragement from profits is not distribution of the results of production on the basis of capital. Distribution is still on the basis of work; it is work that rules distribution under socialism.

The Goal and the Means

The significance of profit in the Soviet Union was underestimated owing to a certain disregard of the law of value. Some Soviet economists incorrectly interpreted that law as an unpleasant leftover from capitalism and said we had to get rid of it as quickly as possible. Shelving the law of value led to arbitrary fixing of planned prices and to prices that operated over too long a period. As a result, prices became divorced from the real value of goods, while profits fluctuated greatly from enterprise to enterprise, even on comparable articles. Under those conditions profits were poor reflections of the actual achievements in production. Because of this, many economists and economic managers began to consider profit as something completely independent of production and, hence, as a poor guide in matters of economic management. This is the delusion many Soviet economists, among them the present author, are now trying to expose. We do not intend to go back to private enterprise but, on the contrary, to permit the economic laws of socialism to operate. Centralized planning is wholly compatible with the initiative of enterprises. This is as far from private enterprise as private enterprise is from feudalism.

The law of value is not a law of capitalism but a law of all commodity production, including planned commodity production under socialism. The difference from capitalism is that the goal and the means have changed places. Under capitalism, profit is the goal, and the satisfaction of the needs of the population is the means. Under socialism it is just the other way around. Satisfaction of the needs of the population is the goal, and profit is the means. The difference is not one of terms but of substance.

"Time" and the Soviet Economy

Soviet economists can only smile when they read how Time interprets the socialist planning system. It says: "A knitwear plant ordered to produce 80,000 caps and sweaters naturally produced only caps: they were smaller and thus cheaper and quicker to make." In other words, the factory had freedom of choice. But elsewhere the same article says that factories are tied hand and foot by the plan, and that the plans account for each nail and electric bulb. Where is the logic?

Another example: "Taxi drivers were put on a bonus system

based on mileage, and soon the Moscow suburbs were full of
empty taxis barreling down the boulevards to fatten their
bonuses." But every Moscow schoolboy knows that the bonus of
taxi drivers is based on the amount they collect in fares. Empty
runs are a disadvantage. As a matter of fact, there is a restric-
tion on the mileage of empty runs. Taxis in Moscow and many
other cities are radio dispatched, the purpose being to reduce
empty runs. Such lack of knowledge on the part of the <u>Time</u> staff
can hardly make for an objective appraisal of the Soviet
economy.

The magazine's statements on more serious matters are
just as informed. Experimental garment factories, it says,
"showed such a resounding improvement in efficiency — and such
'deviationism' — that many Kremlinologists assumed they had
contributed to Nikita's downfall."

In the first place, these factories did not show any "resounding" improvement in efficiency. On the contrary, their
output dropped owing to a greater outlay of labor for more
painstaking manufacture and finishing of the articles. The only
thing they showed is that when given the right to plan their
output on the basis of orders from stores, they can make good
suits of wool and man-made fiber mixtures at a lower price.
Customers readily buy these suits.

In the second place, what kind of "deviationism" is it if the
"deviation" was made in conformity with instructions issued by
the Economic Council of the USSR in March 1964 — without any
direct participation by Professor Liberman, whom the Western
press cites, without having sufficient grounds for doing so, on
every occasion when steps are taken to improve the Soviet
economy. My modest role, like that of many other of our
economists, is to study methods of improving economic management on the basis of the principles and economic laws of
socialism.

Rivers Don't Flow Backward

Soviet economists have no intention of testing the economic
methods of private enterprise. We expect to get along with our
own methods, sharpening the tried and tested instrument of
material incentive on the grindstone of profit. This has been
one of our instruments for a long time, but it has grown dull,
chiefly because we didn't use it enough. Now we are sharpening
it and it will, we hope, serve socialism well. But this does not

mean that we are either giving up a planned economy or turning toward the system of private enterprise. Rivers do not flow backward. And if, at high water, rivers make turns, they are simply cutting better and shorter channels for themselves. They are not looking for a way to go back.